# Design for
# Social Innovation
# in Canada

Lorenzo Imbesi

Afterword by Ezio Manzini

# Design for Social Innovation in Canada

Lorenzo Imbesi

Afterword by Ezio Manzini

COMMON GROUND PUBLISHING 2016

COMMON GROUND

DESIGN PRINCIPLES AND PRACTICES

First published in 2016 in Champaign, Illinois, USA
by Common Ground Publishing LLC
as part of the On Design book imprint
Book imprint editors: Lorenzo Imbesi

Copyright © Lorenzo Imbesi and the chapter authors, 2016

All rights reserved. Apart from fair dealing for the purposes of study, research, criticism or review as permitted under the applicable copyright legislation, no part of this book may be reproduced by any process without written permission from the publisher.

Library of Congress Cataloging-in-Publication Data

Imbesi, Lorenzo.
Design for social innovation in Canada / Lorenzo Imbesi.
pages cm 17.8 x 25.4
ISBN 978-1-61229-810-8 (pbk : alk. paper) ISBN 978-1-61229-811-5 (pdf)
1. Social change--Canada. 2. Community development--Canada. 3. Technological innovations--Social aspects. I. Title.
HM831.I4293 2015
303.4--dc23
2015031749

# TABLE OF CONTENTS

8 | **Introduction & Acknowledgements**

10 | **Designing the Commons**

24 | ## SOCIETY

**Canadian Humanitarian Design** 26 | **The First Electric Wheelchair** 28 | **The PATCH Project** 30 | **BioBlitz** 32 | **Culture.licius** 34 | **Startup Canada** 36 | **Get Volunteering** 38 | **Playing for Keeps** 40 | **Many Hands, One Dream** 42 | **Jane's Walk** 44 | **Coworking** 46 | **The Chantier** 48 | **Freecycle** 50 | **Timeraiser** 52 | **Design Nerds** 54 | **Broken City Lab** 56

58 | ## ENVIRONMENT

**Depave Paradise** 60 | **BIXI** 62 | **Dockside Green** 64 | **Employee Transport** 66 | **Changing Currents** 68 | **S-M-A-R-T Movement** 70 | **Ottawa Good Food Box** 72 | **Not Far From the Tree** 74 | **Rooftop Farm** 76 | **Hidden Harvest** 78 | **Community Urban Food Forest** 80 | **Green City Acres** 82 | **Project Neutral** 84 | **8-80 Cities** 86 | **Strip Appeal** 88 | **Sun Country Highway** 90

92 | ## TECHNOLOGY

**Toronto Tool Library & Makerspace** 94 | **Venio** 96 | **Molecule R** 98 | **Digital Literacy for Women & Youth** 100 | **Shopify** 102 | **Turning The Tide Against Online Spying** 104 | **SurfEasy** 106 | **Fundchange** 108 | **Pain Squad** 110 | **LE WHAF at JUNIPER** 112 | **Minuum** 114 | **Tyze Personal Network** 116 | **Hibe** 118 | **Social Media Tutoring** 120 | **reBOOT** 122 | **Social and Environmental Computers** 124

126 | **Afterword:**
**Design and Social Change:**
**Grassroots Innovation and Community Development**
Ezio Manzini

# INTRODUCTION & AKNOWLEDGEMENTS

This book is dedicated to Design for Social Innovation in Canada, also featuring Ezio Manzini (DESIS Network), in order to understand its multiple expressions and practices. Social innovation is now impacting the interdisciplinary discourse of a number of fields of enquire which design is developing connections: social sciences, business, new technologies, management, and more.

The publication is aimed to better understand social innovation in its interdisciplinary definitions and perspectives, while examining the designers, their approaches and methodologies, the organizations and the ideas, but also the final outcomes from their activities, not only as physical artifacts, but also as services, processes, or community based social systems.

Through a number of case studies of social innovation in Canada collected in the fields of Society, Environment and Technology, we can recognize how the role of the designer cannot be anymore limited to the production of finished objects, rather the creations of enabling tools or platforms for the users/citizens to participate in the project, even allowing forms of personalization. This is a process that may be displayed with the evolution of open source and Web 2.0, where Internet is shifting from the simple function of consultation, to the empowerment of anyone to contribute with individual contents.

Thus, the boundary between the work of design and the involvement of the users/citizens is blurring in a form of cooperation, and the role of the designer evolves to create platforms for interaction and communication, which often reaches the Do-It-Yourself. Self-production and participation express the tactical intelligence of the small solutions, as opposed to the big ultimate ideal project of the modern times.

The projects are including: open source, social enterprises, coevolution and participatory design, design for humanity, co-production, co-design and co-responsibility, crowdfunding, slow food, DIY projects, sharing services and portable enabling technologies.

How can design be an agent for social innovation? What are the conceptual categories emerging? What are the design scenarios and the approaches? What are the outcomes from its activity?

This book is the result of a research project about Design for Social Innovation developed in Canada between 2010 and 2014, while the author was a Faculty and Coordinator of the Master of Design MDes at Carleton University in Ottawa, ON. The eminent interdisciplinary approach of the program can be clearly recognized in the nature of the outcomes of this research.
At that time, Ezio Manzini was involved as Visiting Professor and collaborated with the students in a Seminar: his interactive involvement and contribution have been strategic for the success of this project.

The research was possible thanks to the involvement of a number of students, who collaborated through the years, in particular:
Adam Colavecchia, Alan Cheng, Allan Maltais, Ally Krug, Alyssa Wongkee, Andrew Theobald, Andrew Theobald, Benjamin Fogarty, Brandon Geller, Calvin Ma, Carmen Liu, Chao-Kai Jen, Charles Williams, Christoper Spadal, Cristina D'Alessio, Eduardo Mujica, Emma Peat, Foong Ling Chen, Grace Lee, Heather Jeffery, Hyunji Son, Jamie McFarlane, Jasmine Yeung, Jason Ham, JD Sherman, Jeff Burgers, Jeffrey Lee, Jennifer Bin, Jennifer Vandermeer, John Sherman, Jon Cairns, Katie Roepke, Kelvin Chow, Kenneth Blouin, Kenneth Wiafe, Kyle Lunau, Laura Moor, Laura Palbom, Lena Sitnikova, Leonardo Barros, Linette Brown, Linsday Willcox, Lisa Sleiman, Luis Garcia, Meaghan Mallory, Michelle Zheng, Nate Williams, Nathaniel Hudson, Nathaniel Overtholtzer, Neil Voorneveld, Noha Fouad, Owen Sheedy, Paul Chamandy-Cook, Phil So, Remy Godzisz, Rob Sawatsky, Ruby Hadley, Rylan Smirlies, Sanjeevan Tharmaratnam, Sean Parker, Soo Min Choi , Steve Kellison, Stuart MacMahon, Tiz Cousineau, Victoria Dam, Xinran Fan, Yesmeen Ghader, Zoe Krug.

The publication could not be possible without the support and the encouragement of Common Ground Publisher, in the person of Phillip Kalantzis-Cope and its brilliant team, especially Ian Nelk for patiently steering through the production process.

Special thanks go to Cristina D'Alessio, who has been assisting me and following the entire project since the beginning to the very end of the publication.

Furthermore, it is necessary to mention Viktor Malakuczi and Alfonso Tiberio who worked on the final design of the book.

# DESIGNING THE COMMONS

## Keywords

Distributed Systems; Innovation Communities; Open Design; Co-Creation Platforms; Enabling Solutions; Grassroots

## Abstract

*Designing for Social Innovation means, over all, designing the Commons. The commons is not only a space of resources managed by public or private entities: there is a collective dimension of the common goods which highlights the access and the participation from everybody. On a further note, it is the way in which the asset is 'built' to make it accessible to all interested parties, so becoming a question of design.*

*Goods (both the material and the immaterial ones) should be considered 'commons' not simply because they are free or open, but because they are regulated, nurtured and preserved by an equitable access, through social practices, also taking in consideration the needs of the generations yet to come. Again, it is a question of design.*

*Along with a new notion of 'common good' raised by a class of citizens claiming responses to everyday problems, design may advocate forms of participation in the project, to test tactics of interventions for the public spaces (as environment, but also as culture and technology) which can be provisional, open-source, informal, unsolicited, DIY and micro.*

*At the same time, the democratization of technology and innovation, and the development of the distributed systems is opening a novel chance for designing, diffusing, adopting and scaling local projects into networked infrastructures, and so starting global forms of collaboration and interaction.*

*According to this complex and in-evolving scenario, design for Social Innovation is becoming not a only a new field of action for design, but it can be considered the new 'ethical' background for the design processes of the future, which is changing not only the stakeholders, but also any possible project outcome, whether they are products, systems or services.*

## Introduction

Reflexivity is a category brought up by Ulrich Beck, Anthony Giddens and Scott Lash in order to substantiate the shift in the advanced modernity as a result of reflecting on the consequences of its own actions, such as the environmental threats, the social fragmentation, or the political dissolution (Beck, Giddens, Lash, 1994). The reflexive modernization is a modernization which is starting to reflect on itself, the unintended consequences, the risks and implications of its activities on the entire world. In particular, Ulrich Beck is claiming the emergence of a new risikogesellschaft, which is a risk society assimilating in itself all the dangers and the lateral outcomes from the process of modernization (Beck, 1986). Being aware of the risks, the reflexive modernization works continuously on limiting, reducing and distributing them, which may even end up by reproducing and enlarging the problems.

Hence the current crisis crossing the institutions of society, from finance to politics, to environment, seems to be the outcome of the risikogesellschaft and at the same time also people are realizing to live in a state of hazard, contingency and uncertainty, and so a new demand of change is soliciting for collective projects and bottom-up solutions to real, local and social problems. In one world, for social innovation.

Thanks to technology and networking, people are getting aware of the risks we are facing with the environment and the pollution, the endless local and global wars, the differences in society and unemployment, the collapse of finance. People are also claiming to take care of their common goods and their public space, which is the shared realm related to the environment and the resources, but also the domain of culture and knowledge. An ubiquitous creative intelligence and self-organized actions are mobilising worldwide collective and grass-rooted activities with projects of social innovation, so responding to a wide demand of sustainability and participation in the management of our common goods.

## Our Common Goods

The origin of the commons is to be taken back to the medieval land in England, Wales and Scotland, which was owned collectively by locals as part of the manor for few sharing activities as hunting, fishery, grazing or collecting firewood. Limitations were given according to the quantity and the season to avoid the extensive exploitation and as a consequence the commons was a sort of scheme and a social practice for managing the shared resources. The usage rights were restricted by number and type of animal and by the time of the year, so protecting and sustaining the nature of the place.

The tenure was actually separated from the use, which is somehow what we would currently call the service: the commons are telling about a different relationship between people and goods after the market logic and so the mediation of the public or the private property. Bridging with the analysis of Jeremy Rifkin on the network economy and the information society, the freedom of access prevails over the property developing a more sustainable and collaborative society after consumerism (Rifkin, 2001).

The commons are not just things or resources and it is not about the free and open deployment of the them, but regulating, nurturing and caretaking of goods for an equitable access, use and sustainability through social practices, also for the generations yet to come.

The commons is not a space of resources managed by public, nor private entities: the collective dimension of the common goods is after the ownership, while highlighting the access and the participation of everybody. The common goods can be owned by everybody and by nobody, in the sense that everybody have the right to access and nobody is in the position of claiming any exclusivity. Elinor Ostrom is Nobel Prize for economy and she defines as commons all the material and immaterial resources which are shared, namely which are not exclusive or rival. Rival goods are those which the use by someone may prevent the use by someone else and open a competition (Ostrom, 1990).

Elinor Ostrom wrote a book called 'Governing of the Commons' (1990) where she built a sort of history of the commons reviewing ten centuries of systems of management of communities from the Swiss Alps to the Japanese villages. According to Ostrom, commons is a space, which is difficult to enclosure and anyone may benefit at the same time: as an example air, water and environment are indivisible goods which are difficult to surround and can be enjoyed simultaneously by people. As a consequence, the acknowledgement of a common good is not related per se to the characteristics of the good, but to the social conventions and practices: it is not about another kind of property, but the way it is managed and accessed collectively.

In her work, Elinor Ostrom (1990) validated the theory that the resources can be well managed directly by the communities, so disputing the model coming from Garret Hardin claiming the 'tragedy of the commons' through overconsumption and destruction of the natural resources due to the shared use (Hardin, 1968).

As an example, the Noble Prize displays how a Swiss village is still keeping a medieval community agreement about the use of the land, the water systems, the grazing, but also the maintenance of the forest, the roads and the trails, without the need of privatization, nor of public management. In this case, it seems that the local communities are substantiating a model, which is sustainable and effective at the same time.

While the management of the commons may give a response and a model for the contemporary administration of the resources and an interesting pattern for participation and democracy, the difficulty to delimit clearly what can be considered as common good and at the same time to define the community who should be in charge for the supervision, raises interesting questions and challenges every time. In this regard, David Bollier, a policy strategist and blogger since 1990, in his recent book entitled "Think like a Commoner" (2014) takes the example of a community of surfers on the North Shore of Oahu, in Hawaii. Here it seems to be an incredible place to perform and to prove the talent on the pipelines, but at the same time the really big and massive waves can be a scarce resource and there are risks and dangers. Bollier noticed that sometimes have been conflicts among surfers, especially between locals and outsiders. To deal with these issues, a self-organized collective called Wolfpak worked from their experience around a sort of code to be respected for a surfing etiquette. The big waves can be considered as a common good, such as grazing, and at the same time they raise noteworthy questions: who should have the stewardship for the pipelines? The local authorities or the community of surfers? How can be managed the conflicts between the experience of the locals to protect the place and the right of the outsiders? Whose commons is it? (Bollier, 2014). These are questions which can be easily extended to a number of assets currently involving different people, stakeholders, individuals or institutions.

## Reclaiming the Public Square

On an article entitled 'The Comedy of the Commons' published in 1986, which is responding to Hardin' essay, Carol Rose reckons a number of resources which cannot be recorded as private property, such as the rivers, the lakes, the oceans, the mountains, the forests or the great outdoors (Rose, 1986). Further, she stressed on the fact that having public relations in a public square should be considered a right and a central factor for the social life. The square is the social space where we share, communicate, socialize and develop trust and a social capital, which is essential for participating to the common life. It is interesting that the article from Carol Rose arrived before internet could show its potentials and in a way it envisages the question of a number of common goods related to culture we share after the natural resources, such as the knowledge we developed and stored so far, the language we speak, but also the networks for communication, the technology we developed and even our DNA.
The development of the networks, the democratization of technology through digital innovation and open source culture is making the debate on the management of the commons relevant again, also unlocking few important questions on its potentials and limits. As Elinor Ostrom (1990) validated for the communities in the Swiss forest, networked knowledge cannot be protected just by a private manager, but by the conditions of property use, which must be directly accessible to all the people concerned. It is the way in which the asset is 'built' to make it accessible to all interested parties, so becoming a question of design. The project occurs in defining the geometry of the politics in a radical way, changing the structural characteristics of the interaction and opening to new generative forms of management and cultures of participation.
As Rose claimed the square as a social space to share and a public right for participation, it seems that internet is becoming a highly productive infrastructure for social collaboration and to generate innovation. It is a real

platform for commoning open to the online collaboration of a large number of people which may join their forces to improve and innovate further. Internet seems capable to self-regulate its sharing space without the need of an external system of governance and as a consequence inspired a number of following projects, such as Wikipedia, the system of open-access scholarly journals, or the proliferation of social networks.

Anywhere in the world, people learnt to create and spread online messages, while developing interactive networks. The amount of exchanges and sharing contents are accelerating incrementally, so creating a large space of real-time data and multimedia content generated, distributed and available by and for anyone. The digital and immaterial space of the World Wide Web is acting as a wide public square where a new participatory culture emerges to help people work together to collectively classify, organize, and build information (Delwiche, Henderson, 2013). According to Pierre Lévy, collective intelligence should be considered as the major outcome from internet, which is "a form of universally distributed intelligence, constantly enhanced, coordinated in real time, and resulting in the effective mobilization of skills. […] The basis and goal of collective intelligence is the mutual recognition and enrichment of individuals rather than the cult of fetishized or hypostatized communities." (Lévy, 1997). Collective intelligence happens to be an integrated distributed cognitive system involving people and individuals, but also technology and networks, which means that collective intelligence is anywhere there is humanity networked by technology. Harnessing collective intelligence is crucial for valorising cooperation and participatory culture and can leverage creativity, scientific knowledge, civic engagement, new forms of organization and economy, original collective behaviours and actions, but also to give solution to global scale problems and global crises. The emergence of a collective intelligence and the acquisition of capabilities of self-organization asserts new practices of technological activism, also multiplying the chances and the resources for grass-rooted forms of enterprise and cooperation.

If Internet can be considered as a global public space, the so-called Web 2.0 and the development of the social networks have constituted a new form of civil society. It is a matter of fact that the convergence between Internet and wireless communication is empowering society and individuals, which are now able to produce information and to reach millions of people instantly. The forthcoming development of open data infrastructures, knowledge co-creation platforms, wireless sensor networks, and open hardware may potentially serve collective action and awareness further in the near future.

It is a new wave of collective participation which can be enabled at a scale that was impossible before and this is drawing a number of citizens to be active in being engaged in decision-making processes. Organizations and projects pioneering open democracy are transforming the traditional models of representative democracy and creating opportunities for making meaningful political contributions.

Empowerment means also accretion of knowledge, awareness, transparency and open access in the management of the public sphere.

Manuel Castells identifies this phenomenon as mass self-communication: "it is self-communication because it is self-directed in the elaboration and sending of the message, self-selected in the reception of the message, and self-defined in terms of the formation of the communication space" (Castells, 2001, 2009, 2012) Mass self-communication is based on horizontal interactive networks which can be multimodal and allow the organization of autonomous and alternative projects of communication.

The new social movements can benefit of mass self-communication and develop, share and convey values and objectives worldwide by the social networks, so creating a real opportunity for change. Castells explored the current forms of the social movements emerging from the social networks, which profited of technology for self-organizing and sharing a new political idea (Castells, 2012). While examining radically different political movements, from the riots in the Arab world, to the Indignados in Spain, to Occupy Wall Street in New York, Manuel Castells recognizes a shift from the past in the way the social movements are emerging and organizing themselves: a common trait can be recognized in how the digital space has become an extension of the public square, so creating a short circuit between information, self-awareness and action. Such activist movements were successful because they could spread virally in a networked space by the instant broadcasting of images and ideas: communication became the evidence and the motivation for public indignation, which was stronger than any other ideology coming from the past. At some point, the reflection between the digital space and the physical square were able to develop an extended common field of participation where everyone could be included worldwide.

## Incubating Social Innovation

Manuel Castells (1996, 2001) has been studying the rise of the network society since the beginning and he stressed on the nature of the architecture of the network, which needs to be open, decentralized, distributed, multidirectional and interactive. At the same time, he emphasized on the fact that the users can also be the producers of the technology, while adapting and transforming it to their needs and use. The network society and the collective intelligence is allowing to share innovation in real time, so shrinking the two processes of 'learning by using' and 'producing by using' in 'learning by producing', so activating a virtuous circle between the dissemination and the innovation of technology. The way collective intelligence and technological innovation is happening is not only related with the interaction in the virtual space of internet, but also through physical relations and material contact. The collaborative power of networks is not only about connecting people worldwide with new forms of communication, but connecting knowledge (so enlarging our collective intelligence) and furthermore making interactive links with technologies and objects, which can result in a higher complexity and resilience.

If we are experiencing the new immaterial digital spaces of participation and meeting, at the same time new kinds of physical squares are emerging after the World Wide Web: these are the new incubators and accelerators for supporting grassroots innovation and support social projects to start and to grow their venture. At this point, we are recognizing the proliferation of a number of different innovation hubs, mushrooming in the major cities in the world, such as Fablabs, Makerspaces, Hackerspaces, Living Labs and UrbanLabs. We may check small workshop spaces and co-working environments with digital tools and 3D printing facilities, hosted in University Departments, as in Museums, innovation centres or in grassroots associations. Makerspaces and maker clusters are quickly developing hubs for innovation around the idea of integrating design, manufacturing, engineering and art with the latest prototyping technologies. Interesting projects are emerging at the intersection between open hardware, DIY culture, open source software and open data, while collaborative events such as Barcamps, Hackermeetings, Open Knowledge Festivals and makers Fairs are becoming drivers for sharing innovation, networking and creating collaborative partnerships, workshops and panels. So after the physical space of the co-working hubs and makerspaces, also networking events, fairs, exhibition and initiatives play an important role in promoting projects and social innovation, which may end up in enterprises and grassroots business.

The new scenario brought by the maker movement has just started a revolution in the way we produce physical objects, developing local factories and personal fabricators taking advantage of a worldwide community of peers (Anderson, 2012; Gershenfeld, 2005; Imbesi, 2008, 2009, 2012). If industry is living an historical shift of its role within society and production, open design and distributed microfactories are in the position of independently incorporating all the productive aspects, from the concept, to the manufacturing, to the communication and the distribution, and even the crowdfunding of the project (Imbesi, 2008, 2009, 2012). Even if this is the very beginning of a long process, we can see how the new generations of designers have come to terms with deindustrialization and at the same time are becoming aware of their service and strategic role concerning innovation.

The process of digitalization is leading to the transformation of the nature of the enterprises, while opening to micro-factories and "personal capitalism", able to share locally and globally skills and knowledge, as well as resources and tools, to the accomplishment of projects and products.

In this regard, French social philosopher André Gorz analysed the metamorphosis of labour through the impact of information technology citing Marxism and the dangers of the commodification of human experience, and since 1988 is foreseeing the evolution of the current maker movement and its potentials for innovation and the future: "the division between workers and their reificated work, and between this and its product, is therefore virtually abolished, since the means of production become appropriable and subject to be made in common. The computer seems to be as an universal tool, universally accessible, through which the entire knowledge and every activity can be in common." (Gorz, 1988) Technology is never neutral, but always plays an important role in the processes of social innovation. In particular, the democratization brought up by the information technologies in terms of usability, communication and connectivity, is opening to a wide reorganization of the processes of creation, production and distribution of goods and services, while enabling to forms of re-appropriation of the means of production.

## Co-Producing Solutions

The design of the contemporary commons starts at the convergence of society, environment and technology, all of which seem to be the questions of the present and the future. Design is in the strategic position to start and negotiate satisfactory solutions to meet the common purposes, managing the collective resources and developing the best effective process for a sustainable innovation. The current commons are touching a number of social domains, which are affecting our everyday life, such as health, well-being, inclusion, economy, energy, environment, governance, culture, education, public services.

Design has always played a strategic role in the processes of transformation towards innovation and society: as design has been applied in every sector of industry, to enable the transformation of materials and technology in products, the current shift towards the service society and the knowledge economy has opened to new tasks in the field of the process of management and the user experience. Walking always on the edge of innovation, design is facing the complexity of our times, which the different stakeholders, including engaged communities, are bringing into any process of transformation.

Innovation cannot be anymore considered as a step-by-step practice, but as a complex and dynamic development where people are not considered any longer just as users, but as co-producers and participants involved actively in the different stages of the process.

In 1980 with a famous book entitled 'The Third Wave', Alvin Toffler matched together the consumer with the producer coining a new word: the prosumer is the new actor who can play a brand new leading role in production to meet a new era of mass customization. The concept of prosumerism is actually forestalling the current sharing economy, where the consumers are becoming producers while generating and sharing information, green energy and products realized in 3D printing, often subsidized by crowdfunding. The prosumer dealing with service society and digital networks is opening a hybrid economy, halfway between capitalism and the collaborative commons (Rifkin, 2001, 2011, 2014): transportation, time, knowledge, even tools can be produced, shared and customized. As a consequence, even products and manufacturing can be considered as a sort of service.

Design seems to be at the front line of any user-centered process, considering the complex interaction of any stakeholder within the complexity of systems, products, services, interfaces, hardware and software. At this end, designers can be good facilitators and therefore are able to generate an environment enabling co-creation and participation. Social innovation leverages on the involvement of people for the development of projects and their collaborative interaction as users/co-producers or any other form of participation to the initiative.

Co-design can be intended as a complex and a wide process where individuals and groups can take part interacting and conflicting for the development of projects with a social or community relevance. In this regard, Ezio Manzini recognizes that we are living in a connected world, where sociotechnical systems and infrastructure are drawing distributed networks of food, fabrication, energy, water, culture, and so on (Manzini, 2015). This is opening to new forms of collaborative organization and co-design, where the practice and the culture of design is diffuse and spread: "in our connected world, where everybody interacts with everyone else almost independently of time and distance, this separation of the design team from the rest of the world no longer stands. […] So, in a connected world, all designing processes are in fact co-designing processes […]" (Manzini, 2015). Co-design seems to appear not anymore just as a linear process for the creation of consensus from the stakeholders of a project, rather a difficult, interrelated and contradictory process, which is including a number of different variables to be taken into account. Here the designers and their creativity can take the position of the mediators among the different positions, and the facilitators for the emergence and the development of new ideas and design scenarios. Furthermore, the awareness that every project may affect not just the local everyday life, but may have a larger impact at a larger scale, leverages on the designer's responsibility to respond to the complexity of our natural and social environment and the increase of the number of the fluctuating factors to deal with. Eric Von Hippel in his essay entitled 'Democratizing Innovation' explains the emerging process of open, user-centric and distributed innovation with the development of 'innovation communities', which are "meaning nodes consisting of individuals or firms interconnected by information transfer links which may involve face-to-face, electronic, or other communication" (Von Hippel, 2006). Von Hippel displays that the workspace of such intellectual commons is not just restricted to the development of information products like software, but they are active also in the development of physical artifacts. In fact along with the open software movement, he is bringing as evidence of user innovation in action examples ranging from surgical equipment to surfboards, to kites. Again, here co-design is not only about opening a conversation between different stakeholders, but a real collective contribution in an open process of design, leveraging on the users' abilities to develop new products and services for themselves: "The capability and the information needed to innovate in important ways are in fact widely distributed. […] The net result is and will be to democratize the opportunity to create" (Von Hippel, 2006).

# Scaling the Local Tactics

In his book 'The Practice of Everyday Life' (1990), Michel de Certau described the subtle local tactics of resistance and private practices available to the common man for reclaiming his own autonomy from the all-pervasive forces of commerce, politics, and culture. While examining the microhistories of everyday creativity employed by individuals in negotiating and adapting their environment, he recognizes the consumers as producers able to make living as a subversive art, so developing a distinction between the category of strategy and tactics. While the strategy is tied to a centralized production manufacturing the living environment and dictating the rules for people, on the other side the tactics of consumption are represented through the minute everyday practices and call into question any established and codified operating principle. Furthermore, where the strategy moves within its own confined space, where it is able to clearly scan and control it from the distance, while foreseeing any action in time; on the contrary the tactics are settled by the absence of the ownership of a personal space.

Michel de Certau clearly takes the tactics of living of the obscure heroes of the everyday as a form of local resistance from the global strategies of the centralized forces of production, while developing a narrative from the subjects' individual routines, composed of the habits, constraints and ephemeral inventions, which may have the value of a living project.

Along with a new notion of 'common good' raising by a class of citizens claiming responses to everyday problems as small as the cracks on the sidewalk, design may advocate forms of participation in the project, to test tactics of interventions for the public space which can be provisional, open-source, informal, unsolicited, DIY and micro. The tactics act by developing the characters and the provisional capacity of the local place, discovering its interstices and articulating the organization in an unpredictable way. At the same time, instead of imposing stiff shapes from the top, design tactics may flexibly negotiate innovation at the local level, while encouraging the participation of the community and then acting as catalyst for social revitalization. Design tactics can be considered as an art of arranging the living artificial space to make it habitable by configurations that are never definitive.

While we are experiencing the 'end of grand narratives' and the end of the totalizing visions of history (Lyotard, 1979), at the same time we are acquiring a new awareness of the complex and nonlinear character of any scientific, economic and cultural phenomena which knowledge has to interface with (Morin, 2003). The end of many of the reference points inherited from modernity and the idea of project as an utopian absolute model, is still the source of a new demand of project. Out of the twentieth century big ideologies encompassing a global society, the design tactics go local to give practical solutions and alternatives, while privileging the now and here, rather than an ideal tomorrow yet to come. These may be considered as small islands of heterotopia (Foucault, 1984), where to experience the management of a common good and collective well-being through the experimentation of alternative forms of production, business, work, community.

The democratization of technology and innovation and at the same time the development of the distributed systems is opening a novel chance for diffusing, adopting and scaling local projects into networked infrastructures, and so starting global forms of collaboration and interaction. The possibility to share tools and techniques seems to empower local projects and communities which could be isolated until now, so including them in larger global networks and integrating diverse approaches and practices.

The relationship between the global and the local is becoming more complex and interconnected thanks to communication technologies. In this regard, Anthony McGrew (1992) acknowledges that the transnational networks of goods and information, the movements and the social relations, the practices and the behaviours are expanding and flowing while intensifying the connections and multiplying the networks of the interdependencies. Also Roland Robertson (1992) describes the process of compression of the world in a unified place, which is not standardizing our culture, but multiplying the social relations.

In an interconnected world, the global is becoming the way of being local, so offering a new opportunity to the creative communities to converge in networked systems and then have an influence also at the larger scale. In this regard, Ezio Manzini (2015) claims the emergence of the SLOC scenario, which stands for Small, Local, Open, Connected: "Individually, each adjective and its implications are easily understood, but together they generate a new vision of how a sustainable, networked society could take shape. [...] this SLOC scenario could become a powerful social attractor, capable of triggering, catalysing, and orienting a variety of social actors, innovative processes, and design activities." (Manzini, 2015) Hence, the small and local tactics can be open and connected globally for scaling experiences of social innovation and have a large scale effect in the management of our commons.

## References

Beck, U. (1986) Risikogesellschaft. Auf dem Weg in eine andere Moderne. Frankfurt am Main: Suhrkamp Verlag

Beck, U., Giddens, A., Lash, S. (1994) Reflexive Modernization: Politics, Tradition and Aesthetics in the Modern Social Order. Palo Alto: Stanford University Press.

Bell, D. (1973) The Coming of Post-Industrial Society: A Venture in Social Forecasting. New York: Basic Books.

Bollier, D. (2014) Think Like a Commoner. Gabriola Is, BC: New Society Publishers.

Castells, M. (1996) The Information Age: Economy, Society and Culture. Vol I, The Rise of the Network Society. Oxford: Blackwell.

Castells, M. (2001) Internet Galaxy. Oxford: Oxford University Press.

Castells M. (2009), Communication Power. Oxford: Oxford University Press.

Castells, M. (2012) Networks of Outrage and Hope. Cambridge: Polity Press.

Cipolla, C., Peruccio, P. P. (edited by) (2008) Changing the Change. Design, Visions, Proposals and Tools. Proceedings. Torino: Allemandi.

De Certau, M. (1990) L'Invention du Quotidien. I Arts de faire. Paris: Gallimard.

Foucault, M. (1984) Dits et écrits. Des espaces autres (conférence au Cercle d'études architecturales, 14 mars 1967). In: Architecture, Mouvement, Continuité, n°5, octobre 1984.

Goldenberg, M. (2006) Building Blocks for Strong Communities: Key Findings and Recommendations. Research Report F|58. Imagine Canada and Canadian Policy Research Networks.

Goldenberg, M. (2004) Social Innovation in Canada: How the non-profit sector serves Canadians… and how it can serve them better. Ottawa: Canadian Policy Research Networks.

Gorz, A. (1988) L'immatériel: connaissance, valeur et capital. Paris: Editions Galileé.

Hardin, G. (1968) The Tragedy of the Commons. In: Science, Vol. 162 No. 3859.

Lyotard, J.-F. (1979) La condition postmoderne. Paris: Les Editions de Minuit.

Von Hippel, E. (2006) Democratizing Innovation. Boston: The MIT Press.

Imbesi, L. (2012) Design Comes Out of Industry. New Critical Approaches for Design in the Economy of Post-Production. In: Cumulus Working Papers 27/11, Paris–Sèvres, Publication Series G. Helsinki: Aalto University, School of Arts, Design and Architecture.

Imbesi, Lorenzo (2010) No More Lonely Heroes. From the culture of project to spread Creativity. In: "Design Matters. Designers too. Designers as Human Resources", edited by Cumulus Think Tank Antwerpen: De Boeck.

Imbesi, L. (2009) Networks of Design: critical and social connections between project and self-production. In: "Design Connexity", Conference Proceedings of the 8th European Academy Of Design Conference. The Robert Gordon University, Aberdeen, Scotland.

Imbesi, L. (2008) Design for Self-Production: the Digital Democratization of the Creative Profession. In: "Design & Recherche/Design & Research", Conference Proceedings of Conference Saint Etienne 2008, 6° édition de la Biennale Internationale Design Saint-Étienne. Ecole Supérieure d'Art et Design de Saint-Etienne ESADSE.

Imbesi, L. (2008) Design Power. Design cognitariat at work in the organization of the knowledge capital. In: "Design Thinking: New Challenges for Designers, Managers and Organizations", Conference Proceedings of the International DMI Education Conference. ESSEC Business School, Cergy-Pontoise, Paris.

Imbesi, L. (2008) Etica e Design: Riflessioni. (Tr. Ethics and Design: Reflections). Roma: RDesignPress.

Imbesi, L. (2008) Ethics Become Sexy! A critical approach to Design for the right to access to aesthetics and technology in the knowledge society. In: Cipolla, Carla. Peruccio, Pier Paolo (edited by). 2008. Changing the Change. Design, Visions, Proposals and Tools. Proceedings. Torino: Allemandi.

Jonas, H. (1984) The Imperative of Responsibility: In Search of an Ethics for the Technological Age. Chicago: University of Chicago Press.

Latouche, S. (2007) Petit traité de la décroissance sereine. Paris: Mille et une nuits, département de la Libraire Arthème Fayard.

Lévy, P. (1997) L'intelligence collective: pour une antropologie du cyberspace. Paris: Les Editions de Minuit.

Lyotard, J.-F. (1979) La condition postmoderne. Paris: Les Editions de Minuit.

Margolin, V. (2002) A 'Social Model' of Design: Issues of Practice and Research. In: Design Issues 18:4 (April). Boston: MIT Press.

Manzini, E., Jégou, F. (2003) Sustainable Everyday. Scenarios of Urban Life. Milano: Edizioni Ambiente.

Manzini, E. (2008) New Design Knowledge. In: Cipolla, Carla. Peruccio, Pier Paolo (edited by) (2008) Changing the Change. Design, Visions, Proposals and Tools. Proceedings. Torino: Allemandi.

Manzini, E. (2015) Design, When Everybody Designs. Cambridge, MA: MIT Press

McGrew, A. (1992) A Global Society?. In: Hall, S., Held, D., McGrew, A. (edited by) Modernity and Its Futures. Cambridge: Polity Press.

Morin, E. (2003) L'Humanité de l'humanité - L'identité humaine. Paris: Editions du Seuil.

Ostrom, E. (1990). Governing the commons: the evolution of institutions for collective action. Cambridge New York: Cambridge University Press.

Rifkin, J. (2001) The Age Of Access: The New Culture of Hypercapitalism, Where All of Life is a Paid-For Experience. New York: Putnam.

Rifkin, J. (2011) The Third Industrial Revolution: How Lateral Power Is Transforming Energy, the Economy, and the World. London: Palgrave Macmillan.

Rifkin, J. (2014) The Zero Marginal Cost Society: The internet of things, the collaborative commons, and the eclipse of capitalism. London: Palgrave Macmillan.

Robertson, R. (1992) Globalization, Social Theory and Global Culture. London: Sage.

Rose, C. M. (1986) The Comedy of the Commons: Commerce, Custom, and Inherently Public Property. In: University of Chicago Law Review, Vol. 53 No. 3, Summer 1981.

VV.AA. (2014) Digital Social Innovation. Research Report. European Union D4. Palo Alto, CA: ISSUU.

# SOCIETY

# SOCIETY

| | |
|---|---|
| 26 | Canadian Humanitarian Design |
| 28 | The First Electric Wheelchair |
| 30 | The PATCH Project |
| 32 | BioBlitz |
| 34 | Culture.licius |
| 36 | Startup Canada |
| 38 | Get Volunteering |
| 40 | Playing for Keeps |
| 42 | Many Hands, One Dream |
| 44 | Jane's Walk |
| 46 | Coworking |
| 48 | The Chantier |
| 50 | Freecycle |
| 52 | Timeraiser |
| 54 | Design Nerds |
| 56 | Broken City Lab |

At the time of the knowledge society and the sharing economy, people cannot be considered anymore just as passive clients, but they are an active part of the project as a community or individuals who are aware and interconnected. Furthermore, people are not just people, but they are always detailed by their age, gender, ability, profession, culture, instruction, experience, taste, story, and so on. Besides, people are learning to meet and network through a number of different blends of organizations, private and public associations, non-profit and for-profit enterprises, volunteer programs, which express a wide range of different needs and at the same time claim a new collaborative role in the project and in the process of decision.

Design has the chance respond to a new complexity of society, while producing strategies and solutions to enable, empower and make people being active actors and take a leading role in innovating their own future. The value of design can be found in catalyzing the best resources and in translating them into knowledge and solutions for a wide range of social needs. Here can be found projects for elderly and design for health, volunteers for humanitarian initiatives; community programs for the underprivileged; social networks to promote knowledge on biodiversity; startups fostering multiculturalism and aboriginal culture; grassroots movements to enhance local and small entrepreneurship; platforms connecting social innovators and creative people; projects for place making and local communities; non-profit networks for the sharing economy. After producing our physical realm, design may become a vector of change, while organizing, experimenting and promoting a social economy based on the values of solidarity, equity, and democracy.

# CANADIAN HUMANITARIAN DESIGN
## DesignAid

DESIGNAID

**DATE**

_____

**TIME**

_____

**ROOM #**

_____

To book a room online visit: **http://tiny.cc/rooms**
To cancel or change a booking call:
**Barrie** 705.722.5139 **Orillia** 705.329.3101
**Library UserID required**

**Georgian**

create a social impact in communities in need. It is a social organization, which began with four designers in Vancouver, Canada, who wanted to impact the world in a positive way. Their main focus is "respecting cultural distinction" and "promotion of intercultural learning". DesignAid worked mostly in communities in Africa and Asia. A main project of theirs includes the Bamboo Emergency Medical Transportation Device: the Bambulance. DesignAid impacts locally and globally. The Bambulance is an Emergency Medical Transportation Device which is a sustainable, feasible and affordable solution to getting people to medical centres faster. It is made locally using local bamboo which grows quickly. DesignAid also educates people about design-based community solutions, which enables locals to start thinking of innovative, creative solutions. Through workshops and seminars, DesignAid inspires communities around the world and highlights areas where action needs to be taken.

Additionally, they work with schools to encourage innovation and creativity in children. These are just a couple examples of the collective being humanitarians. Carrying on with their humanitarian framework, their emphasis is on working with and heavily researching the end user in a co-creative and collaborative environment. Secondly, they respect cultural distinction and intelligence by working with local organizations and community members. Overall DesignAid "designs with, not for" rather than trying to become an "instant expert".

Testing and measuring results is a crucial step in DesignAids process reflecting Emily Pilloton's policy of creating and recording measurable results.

Lastly, DesignAid's portfolio of projects is made up of mostly systems, research, development, workshops and studies. They know that the solution is not always (and rarely is) a physical product, and rather may be an intangible arrangement. This reflects the humanitarian design framework's strategy of "designing systems, not stuff", as well as Schumacher's emphasis on designing the immaterial.

# THE FIRST ELECTRIC WHEELCHAIR

George Klein

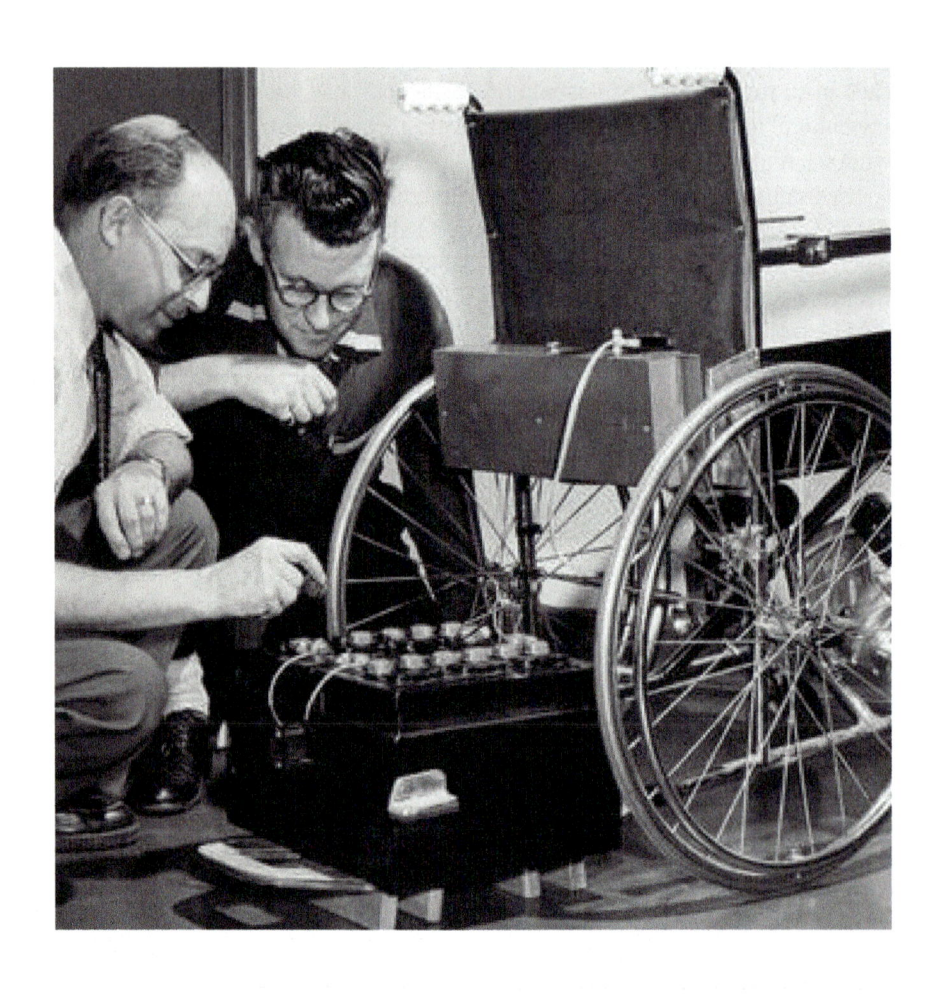

📍 Ottawa, ON

📅 1955

🔑 Universal Design; Co-Creation; Interdisciplinary Design

Since its invention in 1955, the electric wheelchair has significantly changed the lives of those with physical disabilities. Following World War II, there was an increased demand for a wheelchair that could meet the needs injured veterans. George Klein, a Hamilton born designer and engineer recognized this need and while partnered with the National Research Council (NRC) of Canada, developed the world's first electric powered wheelchair. This system was the first to give quadriplegics an independent mode of transportation, vastly improving the users quality of life. A thorough analysis of the hierarchy and specific element of design process demonstrated how, when consciously applied by designers, the hierarchy can help foster the creation and realization of socially responsible designs. There are numerous connections that exist between the Back to Basics analysis and the specific case of George Klein's electric wheelchair. Connections are founded in the consistencies between the design practices implemented in the creation of the wheelchair and the five stages of human need within the hierarchy. Kelin's design approach fulfills each tier of the hierarchy through his design intentions, practices, and processes. The fact that each major design principal touches upon multiple levels of the pyramid, solidifies that socially responsible products can be realized when all fundamental human needs are considered and addressed.

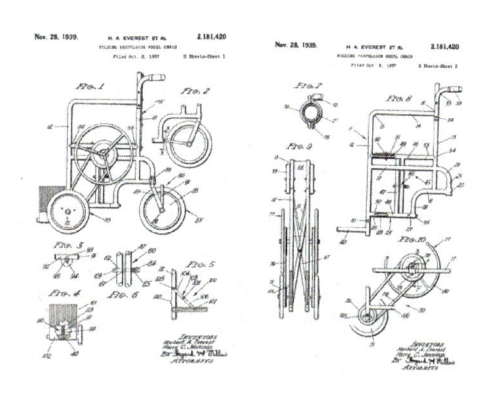

The electric wheelchair was considered to be one of the first principal collaborations between scientists, engineers, patients and doctors. The invention of the electric wheelchair highlighted the need for local co-creation. Today, electric wheelchairs are used on a global scale and have contributed to the betterment of the lives of innumerable people with mobile disabilities. The original electric wheelchair prototype was given to the United States of America to symbolize Canada's commitment to helping wounded war veterans worldwide. Klein's design philosophy expresses his passion for people and the social world through practices and processes, while bringing social innovation to the forefront of the design world, with a very positive local and global impact.

# THE PATCH PROJECT
## STEPS Initiative

📍 Toronto, ON

📅 2013

🔗 http://thepatchproject.org

🏷 Public Art; ARTivist; Urban Communit

The PATCH project stands for Public Art Through Construction Hoarding. Ward 27 councillor Kristyn Wong-Tam, with the help of Sustainable Thinking and Expression on Public Services (STEPS), wanted to make the city of Toronto a more beautiful place by using existing construction hoarding (the fences meant to keep the public safely away from a construction site) as the basis for a public art gallery. The initial project undertaken in 2013 was such a success that it has grown into its own initiative. The PATCH project stages ARTivist events which aim to give local artists the ability to display their artwork in a professional manner, giving them local exposure as well as experience. Proceeds from the PATCH project help to fund community arts programs for underprivileged youth in the area. By displaying the work of local artists, the PATCH project creates a social exchange platform outside the traditional walls of a gallery space. Artwork is brought outside to the public instead of drawing the public in. Going even further, this initiative aims to educate and engage the community through novel events. Youth in the community are given the opportunity to learn artistic techniques through workshops and leadership programs.

Adding art to construction hoarding causes the area to become less of an "eyesore". Where there is more art, less vandalism is typically reported.

The construction company benefits from investing in hoarding art because they no longer need to dedicate time and resources to cleaning up and painting over tags and advertisements. This project can apply to any urban community where there is large-scale construction.

By allowing local artists to gain exposure in their communities and beyond, they are able to work collaboratively with artists in other communities, creating larger scale projects that may have worldwide impact.

# BIOBLITZ
## Royal Ontario Museum, Parks Canada

Ontario BioBlitz

Toronto, ON

2013

www.ontariobioblitz.ca

Biodiversity; Knowledge; Local Ecosystems

Canada's largest urban centre? A local park of 450 volunteers... And biodiversity? That is exactly the premise of BioBlitz. The event is "a special field study, where a group of volunteers conducts an intensive 24-hour biological inventory, attempting to find, identify and record all species in an area". The volunteers consist of curious members of the public as well as biodiversity experts. In 2013, the event took place in Rouge Park, located in the Greater Toronto Area. The comprehensive inventory will be added to a biodiversity database which Parks Canada use to set conservation practices. One objective of the event was to create "an outlet (for scientists) to meet and compare findings" while "inspiring people to appreciate biodiversity at a visceral level so they can spread the word". This illustrates how BioBlitz leverages collective collaboration and social networks to create and disperse new knowledge. Hosted in collaboration with a museum, BioBlitz exemplifies some of the characteristics foreseen to become dominant features in museology. A central focus is engagement, and allowing participants to contribute to real science through a hands-on approach creates very strong connection with the subject matter. Running the event continuously over the 24 hours also introduces a new and exciting learning environment, including after-hours and late-night outdoor activities. The hands-on approach and novel medium is intended to break through our highly digital culture to reintroduce tactile interactions and a love for discovery. The Rouge Park BioBlitz was not the first of its kind, but it was the largest in the world to date, documenting 1,440 individual species. Among them were species never before documented in the park or the area! These were significant discoveries within one small region, and the knowledge will have an impact in helping preserve the local ecosystem.

What if BioBlitz were to spread to other regions? Already, several other ecological areas in the Greater Toronto Area are interested. There are also plans to adopt a five-year rotation among participating sites, enabling scientists to monitor the changes in biodiversity over time. Overall, the knowledge created in one region contributes to the overall understanding of biodiversity and climate change, helping the community get involved in making a difference in the environment in which they live.

# CULTURE.
# LICIUS
## Wendy Chung

Toronto, ON

2012

www.eatlivetravelwrite.com/2013/02/culturelicious-toronto/

Active Aging; Food; Culture

Culture.licious is a startup that hosts intimate cooking events for people looking to learn about different cultures, traditions, foreign ingredients, cooking techniques and recipes. The company is designed to foster a new dynamic of conviviality across generations and cultures. People of all ages and ethnic communities meet in groups of fifteen for cooking classes led by elderly experts with a passion for cooking, who also share their stories and culture. Classes are taught by some of the best home cooks from various ethnic communities around Toronto, hailing from all over the world. Founded by Wendy Chung, Culture.licious aims to preserve the art of home cooking and to bring authentic ethnic cooking into the kitchen, while addressing an emerging social need and a bottom-up approach. The concept of Culture.licious bridges observations from the owner's personal travels, the experience-based knowledge of the elderly, and a city that is open to exploring its own diverse and multicultural roots. The result is a social entrepreneurship that encourages the reintegration of the elderly. Foodies are looking to learn new recipes with ingredients and cooking methods uncommon in North America. Fortunately, there is also an increasing amount of retired immigrants looking to demonstrate their skills to those that are interested. Culture.licious should be considered also a social enterprise with a mission to empower immigrants to pursue their passion while earning a supplementary income. Their Master Cooks are ambassadors of their cultural heritage and take pride in what they are doing. As a result, Culture.licious is a for-profit social enterprise that empowers immigrants, by reintegrating a minority within a minority - elderly immigrants are 'often isolated and harder to reach' due to cultural barriers but are equally willing to contribute. Wendy Chung says: "I believe that by learning about and exploring a different culture, tradition and way of eating endows us with a valuable open-mindedness. That is the beauty of traveling – you connect with people who may further broaden your horizons."

# STARTUP CANADA
## Victoria Lennox, Cyprian Szalankiewicz

Ottawa, ON

2010

www.startupcan.ca

Entrepreneurship; Collaborative Process; Knowledge Economy

DesignAid is a group of designers working together to create a social impact in Startup Canada is a Canadian grassroots, entrepreneur-led movement to enhance the nation's competitiveness and prosperity by supporting and celebrating Canadian entrepreneurship. Their goal is to provide the entrepreneurship community with a strong voice, promote a vibrant entrepreneurial culture and create a unified brand that Canadians can rally around. The strategy is to bring together the collaborative efforts of the entire entrepreneurship community with a clear vision to create more favorable conditions for entrepreneurs to flourish. Entrepreneurs and small businesses are the backbone of the Canadian economy. There are more than 1 million small businesses that employ 48% of Canada's total workforce. Governments around the world are responding to shifts in global markets by strengthening the domestic climate for entrepreneurs as a way to ensure economic growth and prosperity. The launch of Startup America, Startup Britain and similar private-public-civil society initiatives in 2011 have contributed to entrepreneurial growth in more than a dozen countries worldwide. In Canada, the local movement began few years ago. It has been identified that Canadian entrepreneurs are struggling to navigate a complex and highly fragmented enterprise ecosystem, which creates challenges for entrepreneurs to identify and access support, build and leverage national and global networks, and limits the capacity to share knowledge and best practices. An absence of adequate risk capital, combined with a shortage of management and business skills normally fostered through startup failures and second- or third- attempts, contributes to Canada's overall deficiency in entrepreneurial culture. Through a highly collaborative and creative process, communities from coast to coast work together to identify the top challenges facing Canadian entrepreneurs; develop innovative solutions to overcome those challenges; share, build and prioritize ideas, approaches and recommendations for future growth; and ultimately, arrive at a collective vision and strategy that leads to real change and action.

# GET VOLUNTEERING
## Volunteer Canada & Manulife Financial

get volunteering

Ottawa, ON

2010

www.getvolunteering.ca

Collaborative Economy; Active Aging; Volunteering

Get Volunteering is a social program started by Volunteer Canada and Manulife Financial to support volunteerism in Canada. It aims to enhance the spirit of volunteering by bringing people together and getting them involved in the community. The program is largely targeted towards the baby boomers who are entering into retirement. It is designed to highlight the benefits of volunteering as a retirement planning option to boomers, for it "provides them with opportunities to stay active, learn, share and leave a legacy for future generations". The website (GetVolunteering.ca) includes inspiring videos and stories from volunteers across Canada as well as tools and resources to get volunteering. For baby boomers in particular, there is a retirement planning module which discusses how volunteering can it into their lifecycle transitions and considerations such as finance, travel, recreation, family, life-long learning, housing, hobbies, and health. It also includes a volunteer quiz that helps them discover their "volunteer type". For example, a Groupie is one that thrives on being in a group that likes to have fun and get results, and a Juggler is one that enjoys giving their time to a variety of organizations. There is also a competency matrix, which helps boomers determine the skills, competencies, and aptitudes they feel they have to contribute and/or would like to develop by volunteering. As part of Get Volunteering, Volunteer Canada has a volunteer matching tool, an online tool that was developed in collaboration with Manulife Financial and a media solutions to help volunteers and organizations connect.
On the website (Getinvolved.ca), volunteers and organizations get matched with one another based on their interests, skills, and location. The result from the match brings them to a page with all the volunteer opportunities and offers that are available to them. This tool bridges the gap between what volunteers are looking for in a meaningful volunteer experience and what organizations need in terms of volunteer assistance. This social innovation responds to the crisis of the retiring baby boomer population by encouraging the boomers to get involved and contribute to communities with their experience, wisdom, and time. The innovation lies in that it doesn't help the boomers directly, but it helps them give back to the community, and in return again from their community involvement.

# PLAYING FOR KEEPS
## Toronto Community Foundation

Playing FOR KEEPS

📍 Ajax, Hamilton, Toronto, ON

📅 2012

🔗 http://playingforkeeps.ca

🔑 Inclusive Design; Participatory Design; Community Building

Playing for Keeps Neighbourhood Games are locally organized, playful and joyful activities that bring people, neighbours, co-workers, friends and/or family - and even strangers - together to share an experience, have fun and play. Neighbourhood Games cover a full range of activities such as sports, arts, music and food - everything from basketball to hopscotch, from community gardening to grandmas and tot walks, from street dancing to mega glee clubs.

The Games are a catalyst for aligning, leveraging, and focusing existing community assets to enhance the well-being of Ontarians by building healthy active communities and strengthening a sense of belonging.

By working together, people can align social capital objectives with the objectives of the Province of Ontario, the Games organization, participating municipalities, and community organizations. Playing for Keeps is about hitting the ground running, getting the ball rolling, shooting for a better quality of life, fighting childhood obesity, diabetes and heart disease, making the leap toward a healthier, stronger, and better connected community.

Playing for Keeps started leveraging the 2012 Ontario Summer Games towards the 2015 Pan/Parapan American Games to build social capital and create a legacy of healthier, more active, and stronger communities, while nurturing a deepened sense of belonging. We want to leverage physical and cultural play in ways that truly drive enduring impact.

# MANY HANDS, ONE DREAM

Aboriginal Nurses Association of Canada, Assembly of First Nations, Canadian Paediatric Society, First Nations Child and Family Caring Society of Canada, Health Canada, First Nations and Inuit Health Branch, Inuit Tapiriit Kanatami, Indigenous Physicians Association of Canada, Métis National Council, National Association of Friendship Centres, Pauktuutit Inuit Women of Canada

📍 Victoria, BC

📅 2005

🔗 http://www.manyhandsonedream.ca

🔑 National Summit; Health Design; Aboriginal Child

Many Hands, One Dream is a long-term initiative to generate commitment, foster collaboration, and develop and implement solutions that will improve the health of First Nations, Inuit and Métis children and youth in Canada. It is directed by a group of 11 Aboriginal and non-Aboriginal national organizations concerned with the health and well-being of First Nations, Inuit and Métis children and youth. It is started in 2005 in Victoria, BC as a national summit, during this historic gathering, more than 160 Aboriginal and non-Aboriginal health professionals, social workers, educators, parents, administrators, policy makers, researchers and community leaders gathered to initiate sustained, long-term change in the health of First Nations, Inuit and Métis children and youth. Since then, many other organizations and individuals have joined this movement, by supporting the principles of Many Hands, distributing material, participating in other gatherings, and working to improving the health of First Nations, Inuit and Métis children and youth in a collaborative way. Principles of this movement are a new perspective on the health of First Nations, Inuit and Métis children and youth is a document produced following the summit, to articulate the principles that will underpin a new approach to Aboriginal child and youth health in Canada. These principles can be used by individuals, organizations and communities who work with and care about First Nations, Inuit and Métis children and youth. They are: self-determination; intergenerational; non-discrimination; holism; respect for culture and language; shared responsibility for health.

Many hands, One Dream have developed many events and publications about those themes, to raise awareness of public opinion and the Government.

# JANE'S WALK
## Jane Jacobs

Toronto, ON

2007

http://janeswalk.org

Place Making; Community Building; Neighbourhood Development

Jane's Walk is a series of free neighbourhood walking tours that helps put people in touch with where they live and with each other, by fill up social and geographic gaps and creating a space for cities to discover themselves.
The name celebrates the ideas and legacy of urbanist Jane Jacobs by getting people out exploring their neighbourhoods and meeting their neighbours. Jane Jacobs (1916-2006) was an urbanist and activist whose writings championed a fresh, community-based approach to city building, while giving voice to the local residents in neighbourhood planning. Jacobs' wrote incisively and beautifully on the importance of dense and vibrant city-scapes, famously uncovering the 'sidewalk ballet', that intricate dance between neighbours and passers-by that make a street enjoyable and friendly.
Jane's Walks are free, locally organized walking tours, in which people get together to explore, talk about and celebrate their neighbourhoods. Where more traditional tours are a bit like walking lectures, a Jane's Walk is more of a walking conversation. Leaders share their knowledge, but also encourage discussion and participation among the walkers.
A Jane's Walk can focus on almost any aspect of a neighbourhood, and on almost any topic you can think of. Walks can be serious or funny, informative or exploratory; they can look at the history of a place, or at what's happening there right now. Anyone can lead a walk, because everyone is an expert on the place where they live. Since its inception in Toronto in 2007, Jane's Walk has happened in cities across North America, and is growing internationally. Walkable neighbourhoods, urban literacy, cities planned for and by people. Free walking tours held on the first weekend of May each year are led by locals who want to create a space for residents to talk about what matters to them in the places they live and work. Jane's Walk has expanded rapidly. In May of 2013, more than 600 walks were held in over 100 cities in 22 countries worldwide (46 in Canada). Jane Jacobs believed in walkable neighbourhoods, urban literacy, and cities planned for and by people. In her book The Death and Life of Great American Cities, she wrote: "Cities have the capability of providing something for everybody, only because, and only when, they are created by everybody."

# COWORKING
## Centre for Social Innovation

📍 Toronto, ON

📅 2004

🔗 http://socialinnovation.ca/

🏷 Collaborative Economy; Knowledge Economy; Incubator

The Centre for Social Innovation is a social enterprise based in Toronto with a mission to catalyze social innovation in Toronto and around the world and specialized in the creation of shared workspaces for people or organizations with a social mission. While providing its members with the tools to accelerate their success and amplify their impact, it is building a movement of nonprofits, charities, for-profits, entrepreneurs and activists working across sectors and in areas from health and education, to arts and the environment. It has three locations in Toronto, and one in New York City, that serve as shared workspaces, innovation labs and community centers, and where it rents private offices, private desks or shared desks, and meeting and event space, to social innovators and entrepreneurs. The Centre's mission is to catalyze social innovation and to foster collaboration by connecting social innovators and entrepreneurs, and providing them with programming such as workshops, seminars, competitions and mentorship opportunities to accelerate their success. CSI also incubates a limited number of social innovations, providing them with programmatic, strategic, administrative and or financial services. The Centre for Social Innovation, was founded in 2004 by Tonya Surman of the Commons Group, Margie Zeidler of Urbanspace Property Group, Mary Rowe of Ideas That Matter, Pat Tobin of Canadian Heritage and Eric Meerkamper of DECODE. Its first location, CSI Spadina, opened in the Robertson Building owned and operated by the Urbanspace Property Group. In 2010, the Centre for Social Innovation bought a 36,000 sq. ft. building with the help of a financing model called The Community Bond. In 2012, the Centre for Social Innovation opened CSI Regent Park in the Regent Park neighbourhood, Canada's largest public housing community, currently undergoing Canada's largest community revitalization. In 2012, the Centre announced it would open a location in New York City in Manhattan's Starrett-Lehigh Building. In October 2014, the Centre bought the 64,000 sq. ft. Murray Building, located right across the street from CSI Spadina. The Centre purchased the building with 'community bonds,' which are low-interest loans made by private citizens who agree with the centre's mission and want to help it expand.

# THE CHANTIER
## Chantier De L'Économie Sociale

**CHANTIER DE L'ÉCONOMIE SOCIALE**

Montreal, QC

1999

http://www.chantier.qc.ca

Collaborative Economy; Sustainable Development; Equitable Economy; Incubator

The Chantier De L'Économie Sociale is committed to building a plural economy that aims to produce returns for the community and protect the common good, along with the communities' needs and aspirations. Social economy enterprises are collectively controlled, and contribute to ensuring the ongoing economic, social and cultural vitality of communities.

The mission of the Chantier De L'Économie Sociale is to promote the social economy as an integral part of Québec's plural economy, and in so doing, contribute to the democratization of the economy and the emergence of a development model based on the values of solidarity, equity, and transparency. The activities of the Chantier De L'Économie Sociale are about enabling the collaboration of social economy stakeholders and partners at the regional and Québec-wide level; promoting the social economy as a vector of social and economic change; creating conditions and tools that facilitate the consolidation, experimentation, and development of new projects; and participating in the construction of alliances with other socio-economic stakeholders and community action groups that support this development model, also at the international level.

The history of the social economy is the history of women and men who have mobilized to meet the challenges of their times and the needs of their communities. For over 100 years, the social economy has influenced the development of Québec. In a variety of forms and denominations, throughout generations collective organizations and enterprises have contributed to the development of a more human-centred society and economy.

The Chantier De L'Économie Sociale rose out from the same impulse and need. In March 1996, the Groupe de travail sur l'économie sociale (Task force on the social economy) was formed in preparation of the Summit Conference on the Economy and Employment. The task force had six months to assess the potential basin of solidarity in every region of Québec and afterwards it obtained the mandate to continue for another two years. At the end and in order to consolidate the gains of the social economy, the Chantier De L'Économie Sociale became an independent, non-profit society in April 1999.

# FREECYCLE
## Deron Beal

📍 316 Cities in Canada

📅 2003

🔗 http://www.freecycle.org/

🔑 Sharing Economy; Service Design; Sustainable Development

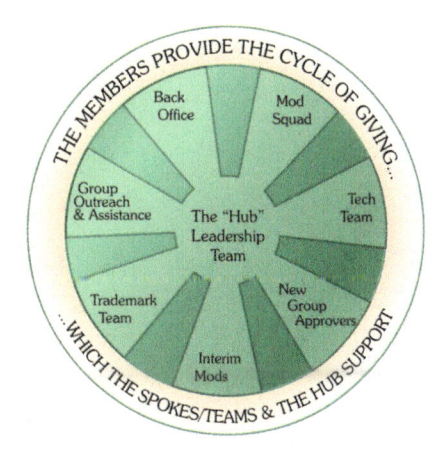

The Freecycle Network is a grassroots and entirely nonprofit movement that organizes a worldwide network of people who are giving (and getting) stuff for free in their own towns and keeping good stuff out of landfills.
It provides a worldwide online registry, and coordinates the creation of local groups and forums for individuals and non-profits to offer and receive free items for reuse or recycling, promoting gift economics as a motivating cultural outlook. The organization originated as a project of RISE, Inc., a nonprofit corporation, to promote waste reduction in Tucson, Arizona. RISE subsequently handed it over to the project leader, Deron Beal and he set up the first Freecycle e-mail group for the citizens of Tucson. The concept has since spread to over 85 countries, with more than 5.000 local groups and 7 million members across the globe.
At the moment, Freecycle is present in about 316 cities in Canada, including Toronto, Ottawa, Vancouver, Calgary and Edmonton. Each local group is moderated by local volunteers and members take part in the worldwide gifting movement that reduces waste, saves precious resources and eases the burden on our landfills. Membership and everything posted is free, legal and appropriate for all ages. As a nonprofit, The Freecycle Network's server and operating expenses are funded by corporate underwriting, on-site advertising, grants and individual donations.

# TIMERAISER
## Framework Timeraiser

**TIMERAISER**
INDEPENDENTLY RUN

Calgary, AL; Edmonton, AL; Halifax, NS; Hamilton, ON; Montreal, QC; Ottawa, ON; Regina, SK; Toronto, ON

2002

http://timeraiser.ca/

Sharing Economy; Knowledge Society; Service Design

Timeraiser is a volunteer matching fair and a silent art auction. The big Timeraiser twist is rather than bid money on artwork, participants bid volunteer hours. Throughout the evening, participants meet with various non-profit organizations in the room to find available skilled volunteer opportunities that meet their needs. Once matches are made, the bidding can begin. Winning bidders have 12 months to complete their pledge in order to bring the artwork home as a reminder of their goodwill. The mission of Timeraiser is to promote civic engagement amongst Canadians in the early stages of their careers; to help volunteer agencies find the skilled people their organizations need and improve the experience of bringing people to causes and causes to people. Timeraiser was conceptualized in 2002 in response to a group of friends wondering how it could be easier to find meaningful, relevant volunteer opportunities. National expansion of the program began in 2006 with the 1st Calgary Timeraiser. A strategy was then developed for further growth, now in 2015 the Timeraiser will be reaching 8 cities across Canada, with several more independently-run events as well.

This expansion was conceptualized with manageable, meaningful growth being a priority – how to grow rather than why to grow. The focus remained on nurturing vital local partnerships, engaging enthusiastic and skilled local volunteers while going deeper into existing cities. We have been able to incorporate our experiences from each city to empower local teams of volunteers, building on our collective knowledge and skills. Timeraiser is directly fostering partnerships with local organizations and empowering local volunteers with the tools, information and opportunity to support to the program.

# DESIGN NERDS
## Vancouver Design Nerds Society (VDNS)

Vancouver, BC

2004

http://vancouver.designnerds.org/

Co-Design; Community Building; Design Thinking

Founded in 2004, the Vancouver Design Nerds Society (VDNS) exist to facilitate, promote and support positive social, environmental and urban transformation by providing a platform for face-to-face creative collaboration. We design, expedite and improve civic engagement processes and techniques that, through the use of dialogue, art, design, media and other creative methods, act as innovative ways to reconnect citizens with their communities and engage them with their environment. As a network of collaborating designers and artists, we share a desire to engage design opportunities with a spirit of creative play and to challenge the normative environment of the city. The diversity of the group enriches the design process and propels discussion far beyond the prescribed parameters of a project, revealing opportunities and unanticipated, often surprising results. Design Nerds are active in architecture, green building, urban design, public art, graphic design, industrial design, engineering, photography, sculpture, policy, sustainability, artificial intelligence, theatre, renewable energy, fashion, dance, film, community development, and so on. The purposes of the Vancouver Design Nerds Society are to: a) Provide space and facilities for interdisciplinary design collaboration and dialog; b) Foster collaboration within the design community through regular organized events and the pursuit of design projects that support socially and environmentally responsive design; c) Engage lay- and expert groups in dialog on design through newspaper, broadcast and web-based media and through participation in public exhibitions and workshops; d) Undertake design projects that engage community and public space; all proceeds of which will be re-invested in the society to further objectives a, b and c above. A Design Jam is an idea factory. It's a fun, fast, creative brainstorming session intended to create a range of diverse visions that address an issue. In the design, architecture and urban planning fields, Jams are employed as one of the many stages of a larger planning, strategy and design process undertaken by the lead design organization and stakeholders involved in a project. By changing the way we think and interact, we change the way we behave and respond to complex problems, therefore creating the social, urban and environmental transformation. The Vancouver Design Nerds Society is comprised of a Board of Directors, a Steering Committee and Community members. The definition of "membership" is still open, but is generally someone who has participated in at least one Design Nerd Jam as well as at least one VDN project.

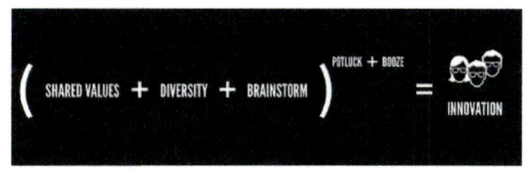

# BROKEN CITY LAB

Justin A. Langlois; Danielle Sabelli; Michelle Soulliere; Joshua Babcock; Cristina Naccarato; Rosina Riccardo; Hiba Abdallah; Kevin Echlin; Sara Howie

**BROKEN CITY LAB**

Windsor, ON

2008

http://www.brokencitylab.org

Place Making; Community Building; Design Thinking

Broken City Lab is an artist-led interdisciplinary collective and non-profit organization working to explore and unfold curiosities around locality, infrastructures, education, and creative practice leading towards civic change. Projects, events, workshops, installations, and interventions offer an injection of disruptive creativity into a situation, surface, place, or community. These projects aim to connect various disciplines through research and social practice, generating works and interventionist tactics that adjust, critique, annotate, and re-imagine the cities that we encounter. Much of the activity has been focused on Windsor, Ontario, a once-collapsing, now gradually stabilizing post-industrial city at the edge of Canada. Their work has been created across media – from temporary interventions to large-scale community events and from gallery exhibitions to various workshops and publications – but they also often take on the role of organizing and facilitating the activity of other artists and creative practitioners through residencies, conferences, and writing projects. Their aim is to creatively respond to the issues directly experienced in a community, while also negotiating the ways in which other community members experience the same issues, differently.

Over the past four years, Broken City Lab has worked with the City of Windsor's Transit Authority to install community-created text-based art in its buses, generated an interactive outdoor projection detailing hundreds ideas for saving the city onto a building in its downtown core, designed and distributed removable micro-gardens, written interactive text-based performance software, told thousands of Windsorites that "you are amazing," projected large-scale messages visible across an international border, hosted artists from across Canada and the US for an interdisciplinary storefront residency project, painted a 350 foot long message on a parking lot, visible from planes and satellites, and led numerous psychogeographic walks, DIY workshops, and community brainstorming sessions in cities across Canada. Their projects and research have been broadly featured in the media, presented and exhibited across North America and have been supported by Canadian Councils, Programs and Foundations. Broken City Lab's work recently appeared in the 13th International Venice Biennial of Architecture as part of the Grounds for Detroit exhibit and the collective was long-listed for the 2012 Sobey Art Award.

# ENVIRONMENT

# ENVIRONMENT

60 | Depave Paradise
62 | BIXI
64 | Dockside Green
66 | Employee Transport
68 | Changing Currents
70 | S-M-A-R-T Movement
72 | Ottawa Good Food Box
74 | Not Far From the Tree
76 | Rooftop Farm
78 | Hidden Harvest
80 | Community Urban Food Forest
82 | Green City Acres
84 | Project Neutral
86 | 8-80 Cities
88 | Strip Appeal
90 | Sun Country Highway

We live in a world with finite resources and we are coming to term with the limits of the planet: the environment is the "common good", which is the resources and the space we share together in our everyday life, but which we inherited from our fathers and we need to pass hand down to our sons.
The environment should be considered the amount of the resources essential for the life of every living species on the planet, without distinction between humans, animals, vegetables or complex organisms. Biodiversity is a value to be protected from destruction and in particular from the action of the man, so to avoid the collapse of the planet. The environment is also the complex habitat we live in, including all the physical, geographical and biological qualities that are allowing us to be alive. Since the urban environment is becoming the anthropic environment par excellence, it seems that every design project will be hosted here while reconsidering the way we live, produce and consume. Here can be found activist projects to restore natural diversity; urban and rooftop farms to promote local food economy; bike sharing and alternative commuting systems; community building on the neighbourhood scale; environment educational programs; distributed food networks and initiatives to reduce food deserts in the urban suburbs; enabling schemes for tracking and mapping the ecological footprint of urban areas; community engagement networks for creating better green public spaces; management solutions for energy saving. The management of the resources is also the administration of the public space and wealth, so becoming a political question. It is not just about saving the planet, but making it available to everyone, so preventing the idea of a sustainable private environment for the few and a contaminated public space for the many.

# DEPAVE PARADISE
## Green Communities Canada

Southern Ontario, ON

www.depave.org

Design Activism; Urban Spaces; Natural Diversity

The initial aim of Depave Paradise project, is to restore natural diversity within the city. Hard, paved surfaces in urban areas prevent rain water from infiltrating the soil and act as heat sinks.

By removing the concrete (depaving) spaces and replacing them with native plants, Depave Paradise allows rain water to easily return to the ground as well as cool and oxygenate the city.

This project originated in Portland, OR, but since then has expanded its roots up into southern Ontario, where 5 Depave Paradise projects are being undertaken. In doing so, Depave is creating opportunities for community involvement, renewing unused urban spaces, as well as a handfull of other natural improvements, such as a higher infiltration rate, preventing toxic chemicals from washing into the storm systems and enhancing the health of urban plants, trees and shrubs. In most North American cities, there is a large trade-off between green spaces and paved spaces; some in which parking spots out number city residents. The growing surface area of parking lots may not have as obvious an impact as the number of cars on the roads, but it is nonetheless a strain to our environment. More parking spaces and roads means a reduced number of plants which can oxygenate the surrounding environment and less open soil to collect rain water and allow for natural infiltration.

Furthermore parked cars leak toxins onto the pavement, which collect and wash away with rainwater down storm sewers. These may feed into lakes and streams, further polluting the environment and overloading the runoff system. Finally, these large dark surfaces absorb heat, creating urban heat islands.

The Depave Paradise project acts as a foundation or process on which urban wildscape projects can be created. In modern cities, the balance between cityscapes and green spaces is heavily one sided towards the city's paved areas. Depave Paradise is a platform to return some green space to the natural environment. Freeing the soil by removing concrete areas creates spaces where rich natural environments can be formed in the middle of the cities. These environments include but are not limited to urban farms and community gardens, parks, green spaces and animal habitats. With the increase of these areas in cities, the urban population can learn to better understand, interact, and co-exist with the natural environment.

# BIXI
## PBSC (Public Bike System Company)

Montreal, QC

2007

http://montreal.bixi.com

Bike Sharing; Collaborative Economy; Urban Sustainability

PBSC (Public Bike System Company) is a bike sharing company that was created by the Montreal Parking Authority in 2007.
The project, which was commissioned by the City of Montreal, aimed to create a durable and concrete solution to reduce the negative effects on the environment and the dependency on cars. The project was modeled after European bike sharing systems, such as Vélib in Paris.
BIXI bikes were officially launched by PBSC on May 12, 2009 in Montreal, with 300 stations and 3000 bikes.
It has since expanded to a number of cities, including Ottawa, Toronto, Boston, London, Melbourne, and Washington D.C. PBSC was changed to PBSC Urban Solutions in 2011. BIXI bikes have also won numerous awards, including a gold Edison award in the "energy and durable development" category, a bronze International Design Excellence Award (IDEA), and a 2010 Good Design Award in the Environment category.
Using a bike sharing program such as BIXI can help to reduce the dependence on cars and oil. By participating in a bike sharing program, commuting time can be reduced, no pollution is created, and the bikes take less space than cars when stowed at stations located throughout the city. BIXI can also contribute to a healthier lifestyle, which, in turn, reduces strain on the healthcare system.
Since Bixi bicycles are shared between many people, they are practically being perpetually reused. This sharing activity has been shown to increase the cycling population of cities in which it has been installed; bicycle frames and some station components are also made of 100% recycled aluminum. In this way they create a continuous path for helping the preservation of the environment.

# DOCKSIDE GREEN
## Windmill West & Vancity Credit Union

DOCKSIDE **GREEN**

Victoria, BC

2001

http://www.docksidegreen.com

Place-Making; Community-Building; Urban Sustainability

Dockside Green is a 1,300,000 square-foot neighborhood development in Victoria BC, that has received extensive accolades for its earth-friendly and people-first design. The defining feature is the integration of residential and commercial real estate, where apartments, shops, and offices exist in close proximity. The project is incredibly successful in implementing efficient technologies on a neighbourhood scale. Because the community was constructed from scratch, the development team had a clean slate to build up from. Starting with a dirt lot allowed these essential technological systems to be woven seamlessly into the neighborhood.

Dockside's global impact has been huge since the beginning of the project but more importantly, the mixed-use community has had a very positive effect on its residents. This development has created a safe and relaxed environment for people to come together as a community in a socially and ecologically responsible way. The community has created opportunities and supported local suppliers. Dockside blends urban living with nature resulting in a high quality of life with minimal impact to the environment.

The great challenge of this project is being able to exploit it even in smaller scaled reality, so as it assumes a retroactive value in systems that still do not have this technology.

# EMPLOYEE TRANSPORT
## Cirque du Soleil

**CIRQUE DU SOLEIL**

Montreal, QC

2006

www.cirquedusoleil.com

Environmental Policy; Social Responsibility; Sustainable Transportation

In the fall of 2006, Cirque du Soleil headquarters in Montreal implemented an environmental policy which aligned and reflected the organizations value for social responsibility. The policy encompassed two main initiatives that encourage and enable employees to conveniently engage in sustainable transportation.
The biking program is multi-dimensional and involves infrastructure and education. The infrastructure consists of a fleet of free self-serviced bicycles, over 200 bicycle parking spots, and a "conference-cycle" allowing 7 people to ride at a time. In 2009, the bikes were signed out 450 times, and throughout the summer the bike racks were almost always full. Free bike repair clinics are put on regularly where professional mechanics are able to fix employees bikes and explain basic maintenance practices to avoid similar problems accruing.
The Public Transit Program is a 50% reimbursement that Cirque du Soleil provides for employees public transit expenses each month if they commit to using public transit for 70% of their travels to and from work. The public transit is infrequent (every 30 minutes) and is less than ideal for getting to work for many employees. With the cost incentive of the Public Transit Program, a total of 300 employees have been encouraged to utilize public transit.
Cirque du Soleil's wide range of sustainable transportation initiatives go beyond compliance with industry standards and exemplify an organization genuinely dedicated to environmental stewardship and social responsibility.

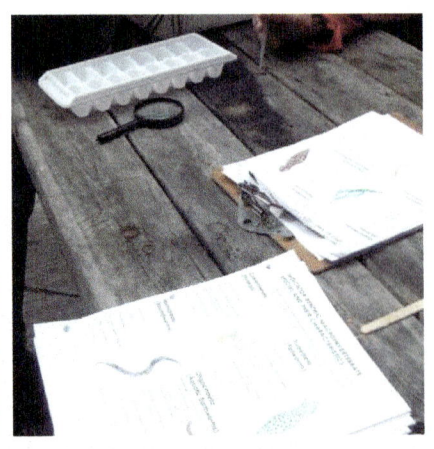

# CHANGING CURRENTS
## Eco Spark

eco spark
discover · act · change

📍 Toronto, ON

📅 2007

🔗 http://www.ecospark.ca/changingcurrents

🏷️ Environmental Education; Water; Crowdsourcing

Since 1995, the Canadian foundation Eco Spark has hosted a diversity of programs aimed at informing youth on current environmental problems such as the overuse of electricity and the pollution of local waterways. Students then put this knowledge to action through hands-on demonstrations and projects. The Big TO River Study is one of the foundations largest programs, and involved more than 2000 Toronto Area teens in water quality testing in 2010 alone. The steps are simple: a teacher is trained in water quality testing using Benthic Macro Invertebrates; the teachers relates the facts to their class; the class takes a field trip to a local stream where they perform test for water quality under the supervision of their teacher and Eco Spark staff member.

Eco quality under the supervision of their teacher and an Eco Spark staff member. Eco Spark provides online tools to analyze the results, as well as a platform for plotting the finding within the region.

These findings can then be used by environmental scientists or legislators to identify and rectify problems with our water management strategies. By crowdsourcing student scientists, valuable data is generated and students are inspired to join the water protection movement.

To continually meet the environmental needs ahead, we must invest in educating the future guardians of our planet's ecosystems. When matters of crisis are brought to the attention of young people, there is an opportunity for them to influence not only their peers, but the generation of their parents, who did not grow up warring about the environmental pressures of today. It is our youth that have the power to disrupt unsustainable patterns and adopt to a more socially aware mindset. Further more, offering the responsibility of being both maker and user of data is empowering because everyone counts. This method of crowdsourcing made accessible, through a digital platform, to allow for these statistics to generate and promote public awareness while supplying governmental organizations with reliable information. Due to the fact that testing the water quality of a stream or river is a laborious process and because Canada has countless waterways, it is beyond the government's capabilities to test all of them. Essentially, Changing Currents is furthering the environmental improvement of our waterways by providing reliable, openly accessible information through the citizen scientists that are educating and inspiring.

# S-M-A-R-T MOVEMENT
## Pollution Probe

**POLLUTION PROBE**
CLEAN AIR. CLEAN WATER

Toronto, ON

2010

http://www.pollutionprobe.org

Sustainable Transportation; Service Design; Social Responsibility

Pollution Probe's S-M-A-R-T Movement is a program that aims to reduce pollution and fuel consumption by helping workplaces reduce their employees' individual car trips. The program is tailored to meet different companies and workplaces to suit their employees' needs. S-M-A-R-T Movement works with companies in the Toronto area to help reduce emissions by using sustainable alternative transportation methods.

S-M-A-R-T Movement is currently working with DuPont Canada, Enbridge Gas Distribution, Exhibition Place, Nelvana, The Town of Markham and Transamerica Life Canada band seven others.

They have helped these companies become more sustainable with everything; from schedule modifications (e.g. flex time, telework, etc.) to public transit (e.g. rapid transit, riding the bus, etc.) to active commuting (e.g. walking, cycling, etc.). S-M-A-R-T alternatives center on solutions for the problems of driving to and from work everyday.

Their messages inform employees that using sustainable transportation will save money and time, reduce stress and improve air quality. Pollution Probe's S-M-A-R-T Movement program includes a coordinator that helps companies strategize customized trip plans. The coordinator also pairs companies with transportation organizations, conducts surveys and provides information sessions to educate employees. The S-M-A-R-T Movement step-by-step manual can be used as an alternative when a coordinator is not available. The manual provides details on designing, implementing and promoting a successful program.

Pollution Probe has continued to provide environmental policy expertise and project management support to Toronto Hydro and smart Canada in support of implementing electric cars.

The S-M-A-R-T Movement is designed to be economical for employees and companies while making the air cleaner. S-M-A-R-T claims to be easy to introduce, implement and sustain for the future.

# OTTAWA GOOD FOOD BOX

## Non-profit community-based initiative

Ottawa, ON

1995

www.ottawagoodfoodbox.ca

Food; Sharing Economy; 0 km

The Good Food Box concept is based on a very successful program developed by Food Share Toronto's Field to Table Program that started in 1994 with only 40 boxes. Now this program delivers approximately 4000 boxes per month in the Toronto Area. This idea was expanded also in many communities from coast to coast of the country that have developed their own version of the Good Food Box, responding to local needs and circumstances.

The Ottawa Good Food Box was developed in 1996 by a group of Community Developers and Community Nutritionists as a way of reaching out to those in the community who were not accessing adequate fresh fruits and vegetables. Basically it is a non-profit community-based initiative bringing neighbors together to buy a variety of good fresh fruits and vegetables at wholesale prices. Goal of this initiative is to purchase food that is in season and is grown as close to home as possible, so to have minimal gas emissions for food transportation.

It works like a large buying club with centralized buying and coordination. And it's open to everyone, so to create a large market and a community of people. There are many points of distribution in the city, especially in the suburb, where there are problems of mobility, especially during the winter period.

Good Food Box can afford to sell at a lower price because the organization relies heavily upon community volunteer support and centralized buying and coordination. Selling at a lower price has been proven a successful way of reaching out to those in the community who cannot access adequate fresh fruits and vegetables. The main difference with the Food Bank is that the Good Food Box is dedicated to selling quality fresh fruits and vegetables at wholesale prices as a means of supporting greater access to low-cost quality food. As a complimentary mission, the Food Bank is dedicated to providing those in need with emergency food assistance.

# NOT FAR FROM THE TREE
## Tides Canada Initiatives Society

Toronto, ON

2008

www.notfarfromthetree.org

Distributed Food Network; Local Community; Sharing Economy

Not Far From the Tree was founded in 2008 to help Toronto homeowners to put their food-bearing trees to good use by picking and sharing the bounty.
It is a Toronto-based fruit picking project inspired by 3 things: the spirit of sharing, the desire to give back to our community, and a passion for environmentally sustainable living. Torontonians with fruit-bearing trees often have fruit to spare: once they register their tree, volunteers will pick their fruit and divvy up the harvest.
When a homeowner can't keep up with the abundant harvest produced by their tree, they let the organization know and they mobilize their volunteers to pick the bounty. The harvest is split three ways: 1/3 is offered to the tree owner, 1/3 is shared among the volunteers, and 1/3 is delivered by bicycle to be donated to food banks, shelters, and community kitchens in the neighbourhood so that the community is putting this existing source of fresh fruit to good use.
This simple act has profound impact. With an incredible crew of volunteers, they are making good use of healthy food, addressing climate change with hands-on community action, and building community by sharing the urban abundance. Approximately 1.5 million pounds of fruit are produced from Toronto's urban canopy each year, and yet most of this goes unpicked. Not Far From The Tree sprang to life in 2008 as a bright and hopeful solution to this surplus of local fruit. During our first harvest season, the project allowed to pick and share over 3,000 pounds of fruit from one neighbourhood with the help of crew of 150 volunteers. Since then, the project harvested over 113,000 pounds of fruit, donated more than 37,000 pounds to over 30 local social service agencies, registered over 1,800 trees for our fruit picking program, and rallied more than 1,800 volunteers to help with the harvest.
Not Far From The Tree was founded by Laura Reinsborough and is a project on Tides Canada's shared platform, which supports on-the-ground efforts to create solutions for the common good. Tides Canada is a national Canadian charity dedicated to a healthy environment, social equity, and economic prosperity. Tides Canada's shared platform provides governance, human resources, financial, and grant management for leading environmental and social projects across Canada, allowing projects to more effectively achieve their missions.

# ROOFTOP FARM
## Lufa Farms

Montreal, QC

2009

www.montreal.lufa.com

Sustainable Agriculture; Hydroponic Cultivation; Distributed Food Network; Sharing Economy

LUFA farms is a rooftop farm in Montreal. Good farm land is disappearing fast. It's either being lost to parking lots and commercial development or it's being slowly poisoned by overuse of synthetic pesticides or herbicides. To make matters worse, good forests are being lost in an effort to make more farm land. Lufa farm is proposing a more sustainable agricultural system by using roofs as fields and hydroponic cultivation method in order to minimize the environmental impact (using a minimal quantity of water to irrigate the plants), while also using biological controls to take care of harmful pests. This keeps the produce free from synthetic pesticides, herbicides, and fungicides.

But Lufa is not just a rooftop farm, it is an interactive way to buy fresh vegetables and seasonal fruits: you can easily select your base basket size from your house and every week and they will pick a variety of fresh seasonal and greenhouse-grown products for your basket. You will be able to customize your order by removing any items that don't suit you and adding as many additional à la carte items as you'd like. You can choose between 100 pick up points in the city of Montreal.

# HIDDEN HARVEST

Hidden Harvest Ottawa (HHO)

Ottawa, ON

2010

http://ottawa.hiddenharvest.ca

Distributed Food Network; Sharing Economy; Local Community

Hidden Harvest Ottawa (HHO) is revealing and sharing the fruit and nuts around Ottawa, in order to make good use of local food and inspire community members to plant trees for tomorrow which will feed and restore the environment. HHO organizes harvests for existing trees and sells edible trees that make sense for Ottawa's future harvests.

Fruit and nuts that would otherwise go to waste on public and private property are rescued by connecting tree owners with those eager to harvest local food. The bounty from harvest events is shared amongst the nearest food agency, the homeowner, the harvesters, and Hidden Harvest Ottawa.

With volunteer support, these harvest events provide the opportunity, education, infrastructure and legal means for people to access the edible fruit and nut trees around. The association facilitates to plant good food-bearing trees by offering high quality trees for sale and for donation to community groups who have vacant land. By removing challenges such as harvesting and planting food-bearing trees, HHO aims to increase our food security, address climate change and evolve our culture to be a food-tree friendly city.

On this regards, in 2013 the Alternatives Journal published a list of Canadian urban foraging organizations across the country that are feeding the needy, while harvesting fruits from the trees in the cities.

# COMMUNITY URBAN FOOD FOREST

## Permaculture Ottawa

Ottawa, ON

2010

http://permacultureottawa.ca

Environmental Education; Sustainable Agriculture; Food Forest

"Permaculture" derives from the combination of permanent agriculture and permanent culture and it is a method to design and to manage the anthropized environment in order to satisfy the need of people as food, fibre and energy and at the same time to preserve the resilience, the richness and the stability of the natural ecosystems. Permaculture Ottawa is a volunteer-based group that promotes permaculture within the Ottawa region, and offers hands-on workshops, panel discussions, information sessions, movie nights, plant and seed swaps, work bees, and other permaculture-related activities.

The Food Forest project differs from a wild forest and an orchard in that it aims to combine elements of the two: it will be a perennial polyculture modeled after a forest ecosystem. The Food Forest will provide ecosystem services (such as erosion control, on-site water retention, wildlife habitat and fodder, soil fertility building, and native species conservation), while also providing useful edible, medicinal, and commercially valuable products for human consumption.

The goal of the Food Forest is to demonstrate that environmental conservation goals can be reconciled with agricultural goals, and that human needs can be met in sustainable ways. This project seeks to adapt innovative ecological agriculture models that are currently being promoted in other parts of North America, and demonstrate the feasibility of the Food Forest model for the Ottawa region.

# GREEN CITY ACRES
## Curtis Stone

**GreenCityAcres**
Sustainable Urban Farming | greencityacres.com

Kelowna, BC

2010

http://www.greencityacres.com

Sustainable Agriculture; Local Community; 0 km

"Green City Acres is a farm with a vision. In 2012, we grew over 50, 000 lbs of food without owning any land, using 100% natural, organic methods and only 80 litres of gasoline. Every year we strive to revolutionize how we farm in order to reshape our local food system to be more environmentally sustainable and socially responsible. We try to change the world one seed at a time".

The mission of Green City Acres (GCA) is to foster social and environmental change through the production of local food, and to help, teach, and empower people to start growing their own. GCA believe that our transition from a petroleum based society is inevitable and how we chose to perceive that potentially devastating event is entirely up to us.

At GCA, they see that as an opportunity to create the world we want to live in, and while we move from a society and food system that is energy intensive, environmentally destructive, and socially inequitable, we can have fun, eat good wholesome food, get some exercise, and reconnect with the soil and our community at the same time.

The GCA products are available in some restaurants or every Saturday at the Kelowna Farmer's Market, so to create a community around this project.

# PROJECT NEUTRAL

## CivicAction's Emerging Leaders Network (ELN)

**PROJECT neutral**

Toronto, ON

2010

www.projectneutral.org

Co-Design; Service Design; Urban Sustainability

In urban areas across Canada, the residential sector represents as much as 60% of greenhouse gas emissions. Worldwide greenhouse gas emissions are expected to increase as more people move to urban areas.

The focus of PN in on established neighbourhoods where the residential building stock is old and inefficient, representing potential energy savings of up to 75% (average of 22%) for households.

Carbon neutrality is an ambitious goal. Yet Project Neutral feels passionately that the most important action that a neighbourhood can take is to start on that journey. With the right leadership, support and tools, PN is confident great things will happen: neighbourhoods will come together and take ownership of their carbon footprint, innovative technologies and funding mechanisms will be piloted, and neighbourhood resiliency will grow in the face of one of the most daunting challenges currently facing our planet. PN knows there is a great deal of concern about climate change and that the average person would like to do more; however individuals are often overwhelmed by the complexity of climate change and frustrated by an inability to link actions to impact, making it hard for them to set priorities. Problematically, energy conservation and greenhouse gas emission reduction targets are rarely translated to the household level. As a result, when it comes to climate change, there is confusion around the model behavior that a household should strive for.

PN uses a community-based social marketing approach to create household level "norms" that make taking action on climate change more achievable, more tangible, and more fun. PN's approach compares an individual household's carbon footprint to others in the same neighbourhood, and to municipal reduction targets. By addressing this disconnect, PN helps individuals set and strive for meaningful objectives.

Project Neutral's model incorporates the following steps:
1 Engage Neighbourhood Leaders;
2 Benchmark Household Greenhouse Gas Emissions and Track Progress;
3 Compare Yourself to Your Neighbours;
4 See What It Takes to Be Carbon Neutral The Household Challenge provides information to help prioritize opportunities and inspire action.

Balancing the amount of 'measurable' carbon produced by the neighbourhood with the amount sequestered or offset. Measurable carbon includes the carbon produced by each household in terms of energy use, water use, solid waste, transportation use and high impact food consumption.

# 8-80 CITIES

Gil Penalosa, Amanda O'Rourke, Jared Kolb, Rafael Vargas, Juliana Berrio

880 cities

Make a Place for People
DUNDONALD PARK, OTTAWA

Toronto, ON

www.8-80cities.org

Service Design; Inclusive Society; Co-Design

8-80 Cities is a group of community engagement experts. Their inclusive approach to engagement allows community stakeholders to directly participate in the process of creating better walking and cycling networks, and more vibrant parks and public spaces. Their community engagement playbook includes a wide variety of interactive and fun activities that are designed to engage and inspire all community stakeholders.

Their final reports effectively summarize key findings and highlight implementable projects that will inspire clients to take immediate action. Part art, part science, 8-80 Cities Community Space Planning service helps clients transform under-used or under-performing parks and public spaces into vibrant destinations. Whether your goal is to create a public living room or action-packed event space, our comprehensive place-making approach can be tailored to projects of all shapes and sizes.

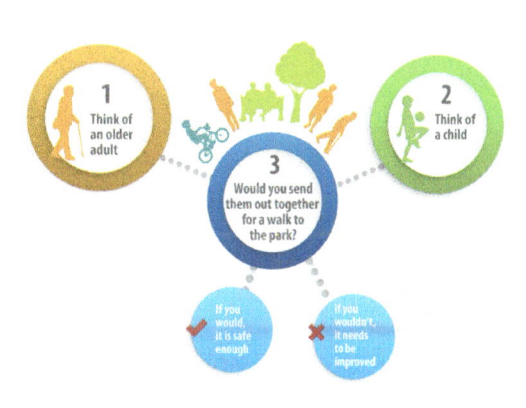

They undertake in-depth quantitative and qualitative research to diagnose problems and develop a holistic strategy to improve the management, uses and activities, infrastructure, and design of the space. Through inclusive engagement processes, their results ensure that all public spaces reflect a community vision and build on local assets.

8-80 Cities helps clients develop integrated mobility strategies that are designed to move people. 8 80 Cities' Community Mobility Planning services include pedestrian safety audits, walkability strategies, cycling plans, and public transportation integration strategies. Their expertise lies in creating accessible multi-modal transportation networks at the community level. Through their ground-up approach, 8-80 Cities determines what the primary barriers are to walking, cycling, and using public transportation, and then works with clients to develop locally-driven strategies that get people moving. By using inclusive community engagement strategies, 8-80 Cities can pinpoint programming and infrastructure that encourage active modes of transportation. The results are happier, healthier, more connected, and more sustainable communities.

# STRIP APPEAL

Rob Shields (Director of the City-Region Studies Centre at the University of Alberta), Merle Patchett (Competition Curator)

📍 Edmonton, AL

📅 2011 - 2012

🔗 www.strip-appeal.com

🔑 Architectural Design; Urban Regeneration; Inclusive Society

In many neighborhoods across North America, small 5-8 store strip-malls, once anchors of local retail activity, have become today's suburban blights: envisioned as community hubs of consumption and services, many of these places are being abandoned, becoming underutilized and dilapidated as the services move out of local neighborhoods in favour of larger-scale shopping districts serving greater catchment areas. Strip Appeal is an ideas design competition, and travelling exhibit, intended to stimulate and showcase creative design proposals for the adaptive reuse of small-scale strip-malls.

The competition asks: how might the small-scale strip be reinvented and redeveloped to local advantage?

With creative thinking and design experimentation, the Strip-Appeal believes there are many ways to transform these ever-present yet ailing built forms to promote walkability, sustainability and community as suburban experience. Strip Appeal exemplifies the commitment of the City-Region Studies Centre at the University of Alberta (CRSC) to promote vibrant and sustainable city-regions in Alberta and elsewhere by engaging communities to research present conditions and future possibilities. This competition provided a unique opportunity to challenge the conventional design of strip malls and to investigate and develop new concepts for sustainable and community-centered suburban architecture for the 21st century.

# SUN COUNTRY HIGHWAY

Sun Country Highway Ltd

## SUN COUNTRY
HIGHWAY

Vancouver, BC

www.suncountryhighway.ca

Energy Saving; Sustainable Resources; Service Design

Sun Country Highway sells the most powerful EV Chargers and the most advanced cord management solutions on the market. Along with our industry leading products, our experienced team offers:
- Expert consultation and advice
- Independent EV testing and equipment validation
- Expertise related to EV infrastructure planning and development
- Comprehensive dealer and preferred installation network
- Research and development.

The vision of the Sun Country Highway is to empowering people to lead global change by adopting a model for economic and environmental sustainability.
The mission is to create the most 'earth-friendly' countries in the world. The principle aim is to empower people to make choices that promote economic and environmental sustainability. The Sun Country Highway want to help green the world's highways by fostering a culture shift toward greener living.
Sun Country Highway provides the safest, most reliable and affordable EVSE chargers on the market, each meeting the rigid testing standards of Underwriter's Laboratories to earn their UL listing. The Sun Country Highway chargers work with all current electric vehicles on the market today.

# TECHNOLOGY

# TECHNOLOGY

| | |
|---|---|
| 94 | Toronto Tool Library & Makerspace |
| 96 | Venio |
| 98 | Molecule R |
| 100 | Digital Literacy for Women & Youth |
| 102 | Shopify |
| 104 | Turning The Tide Against Online Spying |
| 106 | SurfEasy |
| 108 | Fundchange |
| 110 | Pain Squad |
| 112 | LE WHAF at JUNIPER |
| 114 | Minuum |
| 116 | Tyze Personal Network |
| 118 | Hibe |
| 120 | Social Media Tutoring |
| 122 | reBOOT |
| 124 | Social and Environmental Computers |

There is plenty of evidence about the connection between technology and innovation, competitiveness and economic growth. What is more, technology has become open to collective participation at different scales that was impossible before, while empowering and enabling people to have a say in decision-making, collaborative production, in strategically mobilizing and organizing collective resources, in promoting a sustainable change, in responding to community emergencies. Citizens are able to connect each other locally and globally to share common interests and to build projects autonomously, which can become real enterprises and an alternative business. Design is playing a new strategic role in developing processes of co-creation and decentralized platforms through innovation and creativity, where people can generate solutions together for a better future. Here can be found initiatives to share and educate to technology; spaces, services and platforms for the maker movement; crowdfunding platforms for socially innovative projects and charitable causes; DIY kits for enabling scientific knowledge and experiments; online platforms for bottom-up entrepreneurial ventures; systems facilitating collective management and preserving the citizen's privacy; mobile apps helping people aging or healthcaring; devices designed for all; e-waste recycling programs; social organizations for the technological integration and inclusion. Open technologies mean sharing knowledge and awareness, but also coordinating and networking through platforms, which are promoting a brand new participatory culture. Better access to technology, skills and connectivity are enhancing collaborative projects, which are not just solving the everyday problems of people, but furthermore have a political meaning and open a new economy.

# TORONTO TOOL LIBRARY & MAKERSPACE

## Institute for a Resource-Based Economy

Toronto, ON

2013

www.torontotoollibrary.com

Sharing Economy; Maker Movement; Service Design; Crowdfunding

The Toronto Tool Library and Makerspace is a resource sharing project run by the Toronto-based Institute for a Resource-Based Economy (IRBE), and partially financed through the crowdfunding platform Catalyst. The purpose of such a service is to allow the renting of specialized tools, reducing the amount of money and space spent on their purchase and storage. After the initial success of IRBE's Toronto Tool Library, its creators decided to broaden their services to include a 'Makerspace' at a new library location, a space for individuals to exchange information, collaborate on projects, and utilize tools too large or expensive to rent out normally. While the cost of operating the library was fairly small, its membership fees would not have generated enough profits to expand their operational scope in the near future. IRBE sought to use "crowdfunding" to finance its new Makerspace: aggregating small donations from a large group of donors via the Internet in order to realize the project. The Library launched a crowdfunded campaign through Catalyst, a Toronto-based crowdfunding platform focused on socially innovative projects, and was funded successfully in September 2013. Through crowdfunding, the IRBE was able to pitch their idea, outline a plan for realizing that idea, and offer memberships and exclusive training opportunities in exchange for early donations. By incentivizing self-selected donation, they were able to ascertain demand for their project whilst promoting its visibility and bolstering its stated mission within the community. While the community in crowdfunding is composed of "unsophisticated investors", they still find themselves looking for signals of quality in projects, including prior contributions by others, reputation of the founders, and whether or not they care about the proposed outcomes. This project represents collective technological intelligence not only as a collaborative Makerspace, but also by the contribution of crowdfunding. While social media has brought a surge of awareness campaigns and "slacktivism", crowdfunding platforms call on users to collectively assess value through the allocation of scarce resources. This process works to filter support and resources to projects of high quality and meaningful impact, like the IRBE's Makerspace.

# VENIO
## Jon Carr-Harris, Nima Gardideh

Toronto, ON

2012

www.crunchbase.com/organization/venio

Design for Health; Food; User-Centered Design

Venio is a free mobile healthcare app that creates a personalized meal plan based on the user's goals, tastes, dietary restrictions and behaviours. The objective of the app is to help motivated users take control of their health. Managing a healthy balance can be hard and confusing because of the surplus of information available to consumers. There is an overabundance of opinions, food fads, and calorie counters. Venio leverages this by creating an app that synthesizes the data, clarifies it, then translates it into simplified and personalized information. The app does not require a lot of work and time to set up. The user simply inputs the health goals and diet restrictions, and the app will respond accordingly with delicious recipes. The user is also supplemented with itemized grocery lists and nutrition information. The app learns from the user's behaviour and adapts based on cuisine preference and price.

From a business standpoint, Venio's main focus is to build a strong user-base. Venio's initiative is to take time and care in understanding the market in order to present the information in a format enticing to insurance companies. If Venio can demonstrate that the app can be responsible for preventing illness and lowering costs, Venio will be able to monetize on corporate wellness programs. They can leverage existing insurance brokers that already sell to HR departments to develop a health plan for employees within the corporation. For example, every dollar the corporation puts into the wellness program, $4-13 dollars will be reimbursed from the insurance company. Venio also has the option of selling their analytics to restaurants. By predicatively analysing food preference based on geography, they can demonstrate consumer demand. Restaurant owners can use that information to strategically enter the market and set up shop.

# MOLECULE R
## Jonathan Coutu

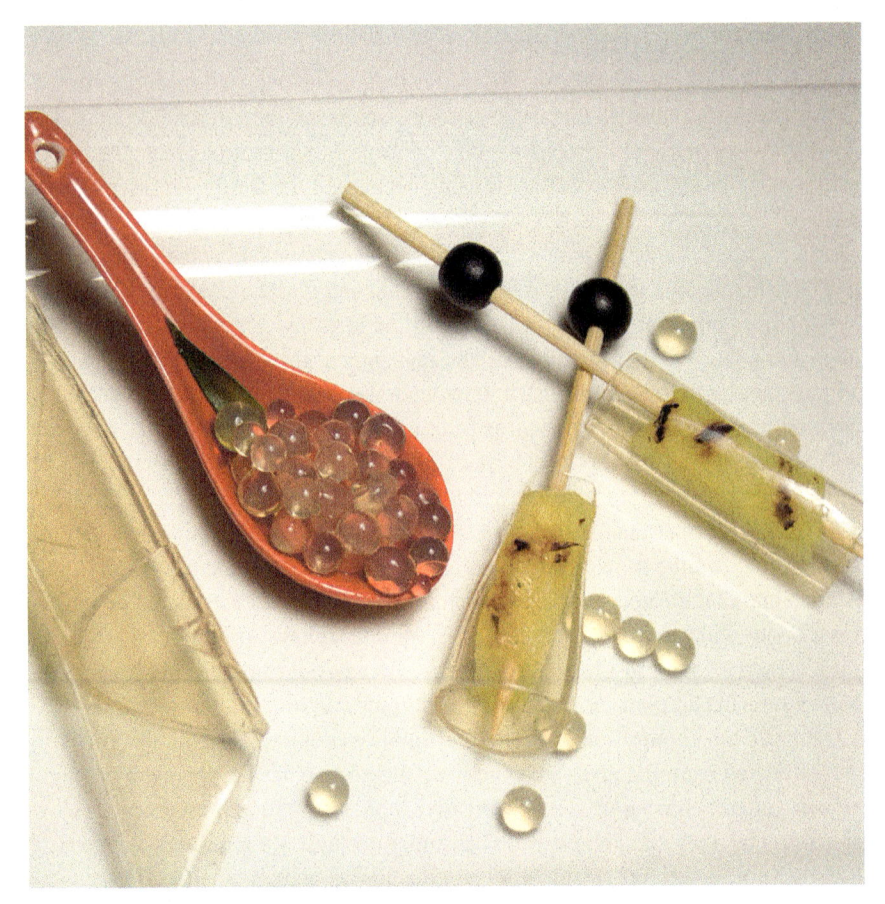

Montreal, QC

2009

https://www.molecule-r.com

Food Design; DIY; Biotech; Open Access

Molecule R is a small company in Montreal that ships molecular gastronomy kits to curious home chefs across North America. Molecular gastronomy is the application of scientific techniques in cooking in order to create new possibilities in cuisine. For the first time, biotech lab techniques can be applied to home cooking, as they have been applied in the past to industrially processed foods and more recently, in cutting edge "Cuisine Moderne" at trendy restaurants. Molecule-R sells DIY kits that include the additives and ingredients, equipment, and training materials to create molecular cuisine in a consumer-friendly, curated experience. Previously, these materials would have to be assembled from a variety of specialty vendors that service science labs, professional kitchens, and industrial food processors. The kits are an example of using design to "humanize" biotechnology and make it more accessible and understandable for everyday people. Molecule-R hopes to capitalize on the trend of DIY biotechnology. The goal is to allow hobby chefs to "hack" and ultimately enjoy their culinary creations rather to make any kind of particular political statement. Nonetheless, this represents a democratization of biotechnology, and of what is typically a very exclusive discourse on the future of food. The crisis that biotechnology presents is it breaks down the definition of "natural" and "unnatural" and hence challenges one of the fundamental ways we understand the world. However today, most of the ways that biotechnology impacts our daily lives is hidden from us and many of the more radical possibilities remain scientific experiments, fringe hobbies, or science fiction. Molecule R on the other hand brings the implications of bio-technology home, literally into our kitchens and dining rooms. Biotechnology has the most immediate potential to changes our daily lives through what we eat. Food is perhaps the most fundamental of human artifacts and the original biotechnology, where we first transformed nature to fit our "taste". Food is a form of cultural expression, with deeply visceral and personal qualities. The debate as to how "natural" food should be, and the exploration of what food could be is about to experience a paradigm shift. This is being accelerated by groups like Molecule R putting the tools of experimentation in the hands of everyday consumers.

# DIGITAL LITERACY FOR WOMEN & YOUTH
## Ladies Learning Code

*ladies
learning
code

Toronto, ON

2011

http://ladieslearningcode.com

Open Access; Knowledge Society; Awareness Networks; Collaborative Economy

Ladies Learning Code programs are designed to help girls see technology in a whole new light. Workshops, camps and after school programs cover a variety of topics from HTML & CSS and Ruby to image editing and blog creation, to 3D printing to hardware hacking with arduinos and more.
LLC is a not-for-profit organization with the mission to be the leading resource for women and youth to become passionate builders - not just consumers - of technology by learning technical skills in a hands-on, social, and collaborative way. In July 2011, it was a small idea - workshops for women who want to learn to code but it quickly grew into so much more. Today, Ladies Learning Code has Chapters across Canada, thriving youth programs called Girls Learning Code and Kids Learning Code as a major force shaping digital literacy education for adults and youth in Canada.

Since the beginning, the founder Heather Payne knew there had to be a better way to learn to code than struggling on her own and she wasn't the only one. The first workshop, an Introduction to JavaScript, was held on August 6th, 2011 and tickets sold out in a day. Now, over four years later, Ladies Learning Code operates in 19 cities across the country and over 10,000 learners attending one of our workshops. Shortly after starting Ladies Learning Code, the organization realized that if they really wanted to have an impact on the number of women in tech, they would need to start younger.

So, in early 2012, they launched Girls Learning Code, which offers workshops, camps and other events for 8- to 13-year-old girls. Hundreds of girls in Toronto have already participated in Girls Learning Code programs, and it is now expanding the program to new cities. A year later, due to overwhelming demand from parents, it was launched Kids Learning Code - co-ed workshops, camps and events for 8- to 13-year-old boys and girls.

One of the best things about the youth programs is the commitment to ensuring their accessibility. Approximately 50% of participants attend Girls Learning Code workshops, camps and events thanks to a full or partial scholarship.
LLC have also made it a priority to acquire laptops so to loan them to youth participants who might not be able to bring their own.

# SHOPIFY
## Tobias Lütke, David Weinland, Scott Lake

Ottawa, ON

2006

http://www.shopify.com

Collaborative Economy, Enabling Solution; Service Design; E-Commerce

Shopify, established by Tobias Lütke, David Weinland and Scott Lake in Ottawa in the year 2006, is an e-commerce platform that allows users to easily create an online shop to sell their products. Shopify shapes the technological landscape by allowing designers the ability to easily create their own online store and access a global market without any previous experience in web development.

The company now has managed to achieve a great deal of success in the last few years, with several offices across Canada and 30,000 users in over 80 countries. The radical technological innovation comes from the establishment of a simple, effective, and versatile tool which allows entrepreneurs the gateway to communicate their ideas, and market them on a global scale.

Creating an e-commerce website with a secure and intuitive platform for customers to purchase their products online traditionally required a company to invest a great deal of time, investment, and resources. Shopify changes this reality by managing all the technical and security features and allowing the user to focus on branding and marketing their products. Users are also provided a strong sense of community through use of blogs where users can interact with each other, company representatives, and Shopify's network of affiliated designers for outsourcing specific design works.

Companies like Shopify have significantly changed the design landscape and how entrepreneurs enter their respective marketplace. What Shopify's services provide is the ability to expedite a products path to market, allowing that product to be vastly marketed globally through the online community.

With this new trend of entrepreneurial technologies, the designer has become empowered with infinite flexibility to bring their ideas into reality. The need for capital expenditures has lessened, the requirement for certain specialized knowledge has declined, and forums to market a product have exploded, meaning it is now a viable reality for anyone to be a designer and confidently compete on a global scale.

# TURNING THE TIDE AGAINST ONLINE SPYING

EFF, CIPPIC, OpenMedia.ca

**PROTECT OUR PRIVACY**

Toronto, ON

2013

https://openmedia.ca/StopSpying

Privacy; Open Democracy; Awareness Network

The Stop Online Spying campaign was a collaborative campaign created in opposition to Canada's Bill C-30, powered by a wide range of academics, public groups and NGOs including the Electronic Frontier foundation, the CIPPIC and OpenMedia.ca. Bill C-30 was legislation proposed by the Canadian government to grant the government unfettered access to the online activities and information of Canadians without the need for a warrant.

The Stop Online Spying campaign's goal was to demolish support for the legislation before it was even put into effect. The strategy to achieve this goal mainly consisted of raising nation-wide awareness about the bill through social networks. They used Twitter as a means to show their unity in opposition to the bill by flooding Safety Minister Vic Toew's Twitter inbox with over 24,000 messages. The campaign also crowdsourced short humorous and meaningful clips to reach an even bigger audience, which were broadcast on TV and YouTube. Finally, an online petition was created which garnered 157,000 signatures. All these strategies contributed to the campaign's success, capturing people's attention and shedding light on the dangers of the Bill. Bill C-30 quickly became widely and vocally opposed by the public, ultimately resulting in the bill being withdrawn in February 2013.

There were several local factors that allowed for the success of this campaign. For example, their Twitter approach relied on the voices of thousands of people working together. The videos they created, and legal expertise were also crowdsourced. By leveraging this they were able to raise awareness about the reality of how the bill compromises the privacy of Canadians. Had this bill been put into effect, the control of the online activities would be handed over to the government, without court oversight.

The government would not only be spying on and controlling one person's privacy, rather the entire society and in turn the Stop Online Spying campaign shows a form of empowerment people and enabling citizens in decision-making processes, while preserving decentralised infrastructures.

# SURFEASY
## Team SurfEasy

Toronto, ON

2012

https://www.surfeasy.com/

Privacy; Big Data; Open Democracy

Most users no longer use the internet from just their homes. With the use of internet in public spaces and on publicly accessible computers, data that travels through these public networks can easily be accessed or monitored before it even reaches the internet, potentially putting the users' personal and private information in jeopardy.

SurfEasy is a simple to use secure web browser that runs directly off of an encrypted USB key. Users only needs to plug in the key, enter their password, and instantly all their web browsing history, bookmarks, usernames and passwords are saved securely onto their encrypted USB key. Users will then be able to enjoy using the internet just like how they would normally with the difference that now they can have peace-of-mind that their online actions and behaviour are no longer being tracked by any third parties that may be of potential threat. The SurfEasy USB key uses custom SSL tunnel encryptions - the same type of encryptions used by major banks - and a highly modified browser to prevent anyone from monitoring what the owner of the key does online. This includes parties such as the user's internet service provider, wi-fi-sniffers, and even company networks. By keeping all internet activity on the USB key and granting access to the internet behind a proxy network, users are able to become "untraceable" as they are no longer leaving any personal information on any computers that they may have used as soon as they remove the key from the USB port. SurfEasy is dedicated to providing their users protection to their online privacy, security and freedom has thus far been a viable solution for many people.

# FUNDCHANGE
## Ideavibes Marketing, Inc.

**fundchange**
sponsored by TELUS

## year one results

**TOTAL IMPACT**

# $105,586

**$55,586**
raised by Fundchange

**$50,000**
matched by Telus

**13** PROJECTS FULLY FUNDED

**54** PROJECTS POSTED

Number of Funders = 251

Ottawa, ON

2010

http://fundchange.com

Collaborative Economy; Open Access; Crowdsourcing

FundChange uses crowdsourcing to raise funds for charitable causes. It is a web application that allows charitable organizations to post descriptions & media and strangers can donate to their cause. The social innovation lies in Fundchange's ability to allow charity to post and receive donations easily and quickly; this opens doors for organizations that do not have the means to launch fundraising campaigns. The Fundchange crowdsourcing model varies from others by regulating the quality and limiting the amount of users. Posting projects is restricted to registered Canadian charities and non-profits, guaranteeing that the projects that are being funded are recognized. This limits the size of Fundchange, making sure that it remains easy to navigate the site, and means that is better at easier to match donor's to appropriate projects.

Fundchange has improved the success of various aspects of social innovation, social entrepreneurships, youth involvement and health and wellness. "Options Mississauga Print and Office Shop," which is operating as a social enterprise that prepares individuals who have an intellectual disability with the skills necessary to gain employment. FarmFolk CityFolk is a not for profit society which targets youth involvement, that works to develop a local sustainable food system. Their projects provide access to and protection of foodlands. FarmFolk CityFolk also supports local growers and producers as well as engaging communities in the celebration of local food, health and wellness, therefore fueling social innovation and change. Le Phare Enfants et Familles is a society that contributes to the well-being of children whose lives are being threatened by an illness. Using zootherapy, children are given the opportunity to interact with animals, with a positive effect on their physical, psychological, cognitive and social well-being. FundChange takes advantage of social media to carry out its goal, such as Twitter, Facebook, or RSS. Feeds and media can be uploaded along with the description to communicate the cause and goals of the organization. For one thing, instant, streamlined fundraising via credit card and PayPal services demonstrate the concept of realtime progression. Additionally, the nature of a web service like this creates more possibilities for target audiences through online communication.

# PAIN SQUAD
## Cundari Toronto

# SickKids®

Toronto, ON

2012

http://www.sickkids.ca/Research/I-OUCH/Pain-Squad-App/index.html

Big Data; Awareness Networks; Open Access

"Pain Squad" is a mobile app designed by Toronto-based agency Cundari to help young cancer patients get better by collecting useful data in the form of a game. A detailed pain journal written by the patient is a necessary part of the cancer treatment process. Unfortunately, cancer patients who have just undergone chemotherapy will feel too weak and discouraged to fill a journal using pen and paper. This is where the "Pain Squad" app plays a big role in making it easy for the young patients at the Toronto Sick Kids Hospital to fill out a detailed journal every day. In the beginning, the game simulates that the young patient is enlisted as a part of Pain Squad, a special police force dedicated to ridding the world of pain. Twice a day, the child receives a dispatch from "headquarters" informing them to file a report. The pain squad then ranks the young recruit by the number of reports they file consecutively.

The Cundari team began the design process by brainstorming ideas to engage the users while providing a platform for doctors to share critical data. Prototypes of the app were used for user testing to gain feedback from the target audience. The detective theme was very well received by the children and the theme was further developed into a dedicated Pain Squad. The app eventually adapted game mechanics of promotion and awards to improve the user experience.

The team made the story more compelling by having selected cast of Canada's top-rated police dramas involved in the announcement of the patient's rank promotion. The project succeeded and struck a chord with everyone. The pain squad app is looking into expanding its usage in 4 other hospital in Canada, and it will be soon available everywhere due to its success.

# LE WHAF AT JUNIPER

## Marc Bretillot & David Edwards

le laboratoire

Ottawa, ON

2012

http://www.lelaboratoire.org

Food Design; Molecular Gastronomy; Experience Design

At Juniper Kitchen and Wine bar in Westboro Village, Ottawa, guests can sample a trending way to enjoy their meal by inhaling it. To do this, chefs are using a new culinary tool called "Le Whaf". This product takes the form of an organically formed decanter, and transforms liquids into vapours using ultrasonic vibrations. The resulting vapour can be poured into a glass for sniffing. With a different version of Le Whaf, the user can actually immerse their face into the tasty haze.
Juniper is one of the first places in North America to share this technology-inspired fare at no cost to its customers. Chef Norman Aitken compares breathing in the vapour to wine tasting, in that you experience aromas with your palate and sinus without having to chew and thus criticize the texture of the food as well. Popular flavours of the vapour are beef consommé and chocolate cake. The practice of "eating clouds" has been highly popular in Europe, particularly in France where the product was developed. Many people use it as a dieting tool to curb their appetite, since the product has the same effect on sensory organs but without any calories. Marc Bretillot and David Edwards are the creators of Le Whaf. Bretillot is a food designer, and Edwards is a scientists. Edwards had inspiration for Le Whaf while working on a new kind of inhaler for pulmonary disorders at a biotechnology firm.
There was another very similar product made called "Le Whif", developed by the same duo. It was created for the very purpose of dieting, so that people could indulge in cakes and sweets.
In the future, our diets will not be the same as they are now. The world population is growing to fast for our agriculture industry to support. People will eventually have to choose alternatives, especially for animal protein products. This will push society to breakaway from traditional culinary recipes in order to embrace new foods. Molecular Gastronomy at Juniper is familiarizing people to new ideas about how, why and what we eat.
It is a glimpse of our food culture future.

# MINUUM
## Will Walmsley

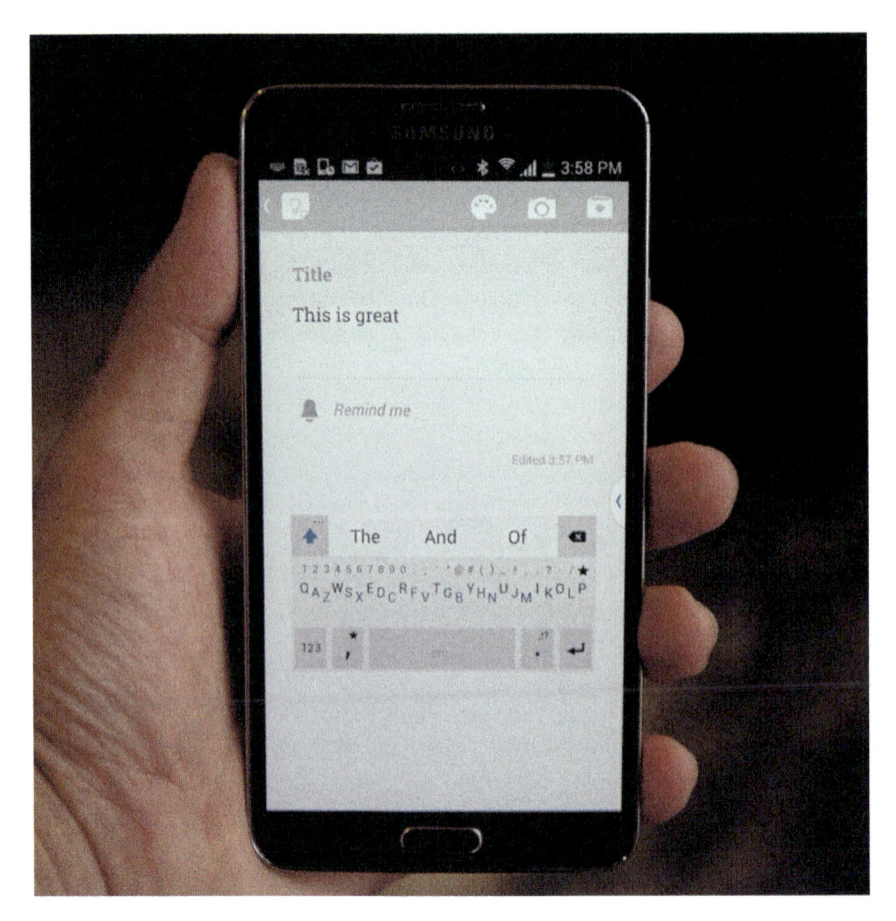

📍 Toronto, ON

📅 2013

🔗 http://minuum.com

🏷️ Interaction Design; Wearable Technology; Open Access; Design for All

Minuum is a simplified touchscreen keyboard designed to take up minimal screen space while enhancing the users's speed and accuracy in typing. The project was initially inspired by a University of Toronto research project: "Invent a better way of typing on touchscreen mobile phones without looking". This in turn encouraged the group of students to extend their knowledge on device tilting techniques which rely on user motion, a common interest in the design of wearable technology. The impact of this project quickly expanded from local to global due to its popular demand.

The interface consists of a tiny, one-dimensional keyboard with user motion technology. By restructuring the typewriter-style keyboard, Minuum allows for a seamless typing experience by adapting the keyboard to a single dimension. The single-line keyboard resolves a common issue of not having enough screen space on touch phones and tablets. Another common problem with traditional keyboards is spelling mistakes and inaccuracy of autocorrect. Minuum addresses this issue by using a specialized auto-correction algorithm that allows highly imprecise typing. The algorithm works in real time in order to interpret what the user types in comparison to what they mean, getting it right every time.

The application of this technology goes beyond its basic use for touch screen products such as phones and tablets. With Minuum, the user can be given the choice to write on any surface, or no surface at all. That is done by measuring the orientation of hand gestures in the air; using an accelerometer/gyroscope in a ring, watch, armband, or handheld device, or with a camera capturing the hand from something like a Kinect, Leap Motion Controller, or Google glasses. With only a single dimension to measure, Minuum makes it easier to apply this technology in these kinds of systems. The Minuum keyboard opens up unlimited potential for applications and uses, especially in wearable technology.

# TYZE PERSONAL NETWORK
## Vickie Cammack, Saint Elizabeth Health Care

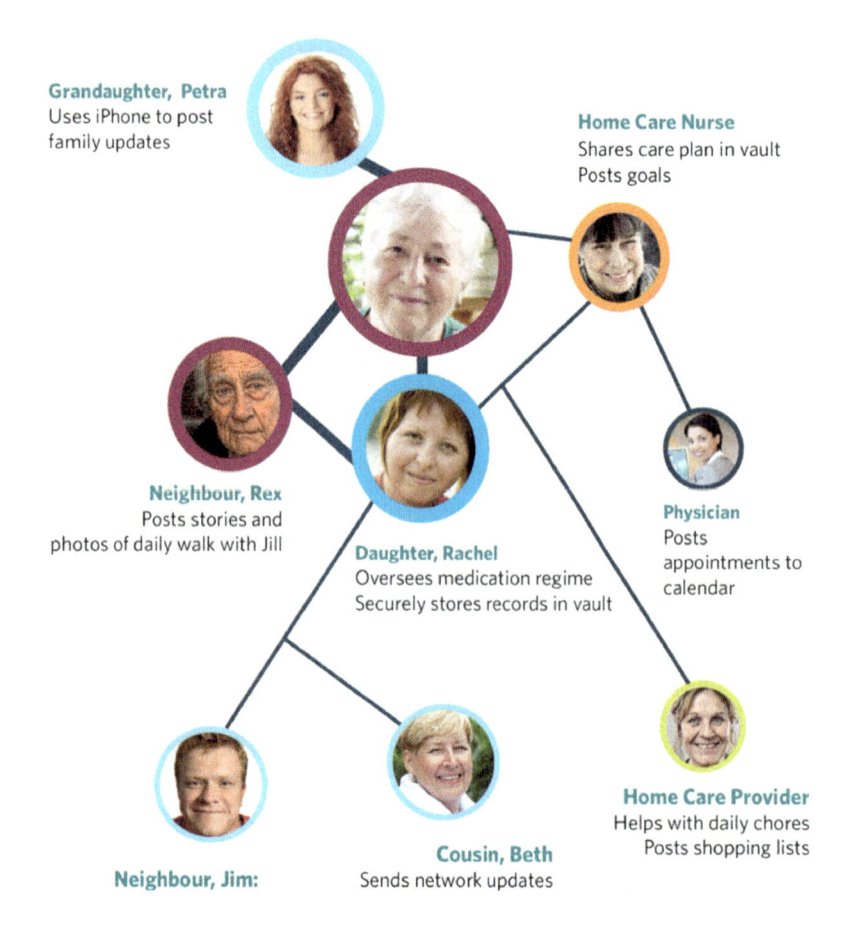

**tyze**

**Grandaughter, Petra**
Uses iPhone to post family updates

**Home Care Nurse**
Shares care plan in vault
Posts goals

**Neighbour, Rex**
Posts stories and photos of daily walk with Jill

**Daughter, Rachel**
Oversees medication regime
Securely stores records in vault

**Physician**
Posts appointments to calendar

**Neighbour, Jim:**

**Cousin, Beth**
Sends network updates

**Home Care Provider**
Helps with daily chores
Posts shopping lists

Markham, ON

2011

http://tyze.com/

Design for Healthcare; Design for Elderly; Social Media; Knowledge Society

Tyze is a private community centred around one person: individuals, families, friends, neighbors and care professionals use Tyze to work together.
Tyze leverages cloud computing to help people care for others. A Tyze personal network is a secure, practical, web-based solution that helps connect people around someone receiving care. Tyze allows to communicate privately with family, friends and helpers, schedule appointments and events on a shared calendar, share files, photos, updates and much more anywhere, anytime.
The radical idea behind Tyze is to create a meaningful and active role for all the people who are a critical part of the care equation. Share vital health information with the people who love the individual who's facing a challenge.
Connect people to each other and to good information. By nurturing the connections, relationships and activities within a person's natural support network, we lay the foundation for their care.
Friends and family have a role and expertise that services could never fill. Of course there is a role for professional services but our personal networks provide a passion, a care and a love that even the most well-intentioned and well designed intervention could never compete with. Tyze brings everyone on to the same page to create the best possible outcomes and to highlight reciprocity, exchange and meaning.
Tyze Personal Networks is a social venture that uses technology to engage, connect, and inform the individual and their personal network members to co-create the best outcomes. Tyze is a social innovation that is using technology as the vehicle for change and was created to contribute to a shift in health and social care from individual to network models of care.
It is a technology-enabled service that creates secure, online personal networks to facilitate the interweaving of relationships and connections. It combines expertise in creating resilient personal support with online social networking technologies. Access to a Tyze personal network is controlled by the patient/client or close family member, or friend, and network members must be invited to participate. Currently, Tyze has 7000 users and has worked with 50 organizations in Canada, the United States, and the United Kingdom. Tyze partners include: the Province of British Columbia, the Robert Wood Johnson Foundation, the United Kingdom Department of Health, and the J.W. McConnell Family Foundation.

# HIBE
## Jean Dobey

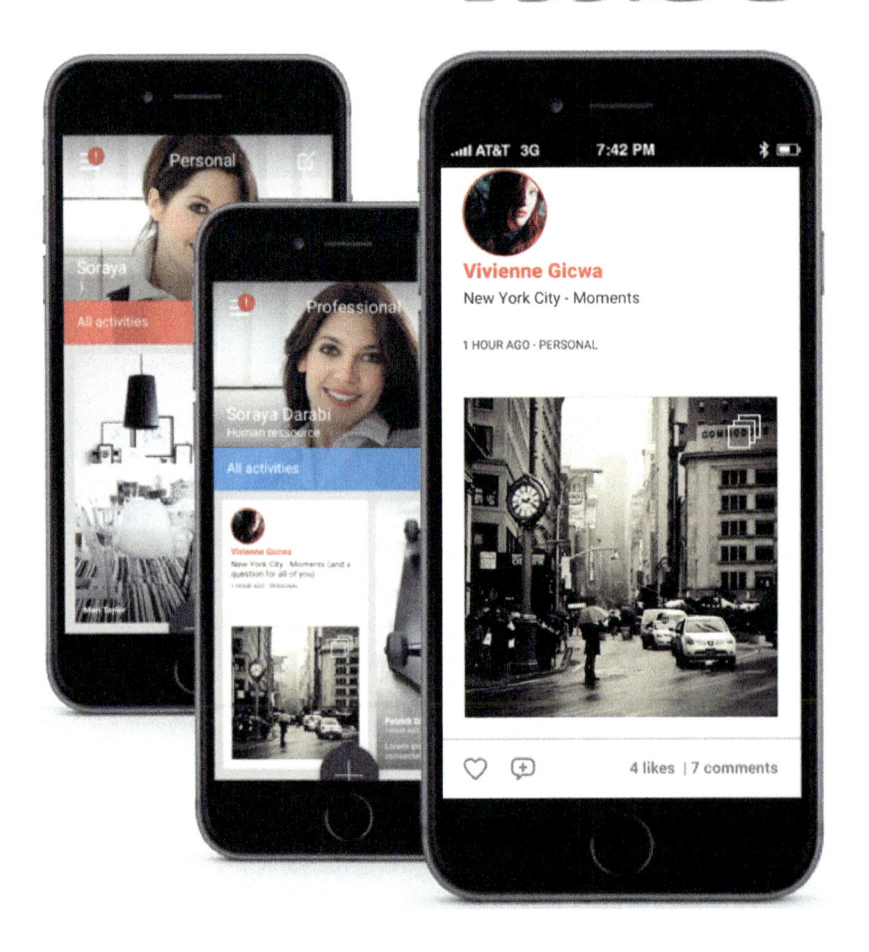

Montreal, QB

2009

www.hibe.com

Social Network; Privacy; Open Access

Since 2009, Hibe has been developing a social networking platform designed to give its users more control over their online identities by allowing them to customize and filter what content is shared to their contacts. The Montreal company has placed a higher priority on the user's fundamental right to privacy - members own all of their posted information, have total control over who is able to view your content, and does not sell your information to advertisers.
On Hibe, users use booklets and facets to share and organize their content. Booklets are interest-based spaces where users place their content to be shared among their chosen facets - groups of organized contacts such as friends, family, co-workers, or the public which are compartmentalized from one another.
Hibe's solution encourages a user to 'Be Yourself', which they believe is possible by reinforcing their fundamental expectations of privacy.
Their manifesto and approach are a response to the crisis of privacy and identity in our social realm. Hibe aims to hinder the unwanted and increasing social convergence that occurs today when we use social networks and other tools. It is an alternative and countermeasure to options that, while popular, may not do enough to protect the privacy and identities of users, or may even believe that "life sharing" is enabled through the erosion of privacy and advent of openness. It facilitates greater impression management in the digital realm, of social identities that are already kept separate in real life.
People can enjoy the liberties and privileges offered by the different norms of various microcultures without concern that social convergence will force the resolution of their multiple identities.
Facebook is a powerful, paradigm shifting tool that offers users richer and easier connectedness, communication, sharing, but there is an almost forgiving, common acceptance of its flaws and the resulting undesired outcomes. Users seem to be acquiescing, adjusting, and augmenting its presence while sacrificing their privacy. People will be interested in technology that enhances their identities. Hibe has the potential to do this by keeping your identities quite separate in its network, just like you would in disparate social contexts in real life.

# SOCIAL MEDIA TUTORING
## Anthony Quinn

Toronto, ON

2011

www.carp.ca

Open Access; Service Design; Active Aging

Intergenerational Social Media Tutoring is a unique collaborative learning service, partnered with CARP (A New Vision of Aging for Canadians) and Thornlea Secondary Secondary, as well as Maple High School. This pilot program has occurred twice at Thornlea Secondary School and once at Maple High School. Both schools are located in the Greater Toronto area.
The opportunity was initially launched at Thornlea Secondary School in spring 2011 and involved the teenagers at the school's personalized education program. Intergenerational Social Media Tutoring is a program whereby seniors attend their local high school for four to six weeks and learn basic computing skills such as Facebook, Skype, YouTube and Twitter. For each lesson a student is chosen to lead the session on the topic and prior to the session the student is expected to design a lesson that is interactive and informative. Following each lesson, students and seniors are matched to work together slowly through the assigned topic. The students benefit from the interactions with the Boomers; whereas, the Baby Boomers gain knowledge from the students.
Through this interactive experience, students boost their self-confidence and learn from their experience.
CARP has recognized an opportunity in teaching Baby Boomers and seniors important computer skills that they may not have acquired. As well, they have acknowledged the opportunity in providing students with a volunteer experience where they can utilize their expertise. The program outlines design criteria that must be met in order for a Secondary School to join the program. Interested schools must have a school computer room (or two) and an adjacent classroom for a few hours after school, once a week for six weeks. In addition, a ratio of one student for every two seniors is required along with the supervision of a teacher. Last, access to one computer per senior with internet access to social media sites is mandatory. Furthermore, each lesson is thoughtfully designed. The teacher and students spend a session planning the curriculum and polishing their presenting skills. A poll of the ability and interests of the group is always taken to ensure everyone is benefiting from the lesson. Moreover, if a gap exists between the abilities within the group, the group is divided into smaller groups. Although this program is still in the start-up phase, it has proven to be a successful collaborative learning experience. The program is continuing to encourage CARP Chapters across the country to replicate the service within their local communities. Globally, the program can be replicated to provide collaboration worldwide where generations learn from each other.

# REBOOT
## Collin Webste

reBOOT canada

Toronto, ON

1995

www.rebootcanada.ca

Open Access; E-Waste; Awareness Networks; Active Aging

reBOOT Canada was founded in 1995 by Collin Webster. It is a Canadian registered charity that seeks to provide under-served Canadians with access to current computer technology. Their reCYCLE program refurbishes donated computer hardware and distributes to organizations across Canada through reCONNECT. reCONNECT was introduced in May 2008, and aims to improve the lives of Canadian senior by advancing their computer skills and allowing them to connect with their families, friends and communities. This is achieved by offering seniors at retirement homes and senior centers free access to computer labs which are equipped with hardware, software, internet access, volunteers and computer technicians.

The reCONNECT program is only made possible through the contributions of our community. Companies, schools, and individual donations provide the resources that enables reCONNECT to impact our community. Companies provide hardware to refurbish, software, volunteers, and cash sponsorships. Many of these supporters are locally based companies as they have better understanding of local needs. However, reBOOT is open to all partnerships whether regional or international. Schools provide a mutually benefiting co-op program for its students, and finally, individual donations allow more hardware to be refurbished and redistributed. Our community at large is impacted by the work of reBOOT, besides seniors, their families and friends are affected as they are able to maintain healthy relationships. The environmental health of our society is benefited as e-waste is diverted from the landfill as reBOOT follows sustainable environmental stewardship in handling obsolete technology equipment. The local nature of reCONNECT allows this program to create such impact in our society, as it uses locally sourced used electronics, local knowledge and skills to empower local citizens. To extend global impact, global firms can follow their model and provide similar service.

Another approach for reBOOT Canada is to involve more international partners to expand this system to different countries and their communities. In this way branches of international firms can cooperate with reBOOT and refurbish and redistribute locally, so these programs can be far reaching to developing nations.

# SOCIAL AND ENVIRONMENTAL COMPUTERS

Insertech

*insertech* ANGUS
Service de formation

Montreal, QC

1998

www.insertech.ca

Open Access; E-Waste; Awareness Networks; Collaborative Economy

Insertech Angus is an IT social enterprise, which can be considered as an example of sustainable development in action. Its non-profit social integration model produces quality products and services while promoting social development and achieving the highest environmental certifications, while its activities reduce the environmental impact of IT equipment while providing a route to social inclusion to those that might otherwise be lacking.

The mission of Insertech is the socio-professional reintegration of young adults and recent immigrants in Québec society. This mission is achieved through paid internships in information technology and personalized social, cultural and professional training. Insertech's mandate also includes the promotion of sustainable development through the harmonization of the economic, environmental and social aspects of its activities.

Insertech Angus started in 1998, under the name Cifer Angus, in the wake of the development project called Technopole Angus. The Angus Development Corporation and the Rosemont-Petite-Patrie had then associated at Rosemont College and the Montreal School Board to set up a new social integration and economic resource for young adults in difficulty.

Since then, the small company grew up and now Insertech continues to recover computers, repackage them and sell them at low cost to citizens and social organizations. However, its economic activities have evolved, becoming much more technological while emphasizing environmental concerns. The training program has been enhanced to accommodate the needs of youth, new tasks within the company and the evolution of the labor market.

Insertech provides customized guidance and a six-month salaried socioprofessional integration program to young adults that are struggling to find their places in the labour market. A charity registered with the Canada Revenue Agency, Insertech is a member of the Collectif des entreprises d'insertion du Québec and works in partnership with Emploi-Québec.

Further, Insertech promote the responsible reuse of IT equipment extending the useful life of IT devices by refurbishing and repairing them so that they can be reused by the community. What cannot be reused is recycled locally in an environmentally friendly manner. The focus on reuse and repair is kinder to the environment than direct, immediate recycling.

Finally, Insertech is serving the community, while selling refurbished IT devices and technical services to the locals in a way that minimizes environmental damage. This is allowing individuals and organizations access to technology at a low cost and in a responsible manner.

# AFTERWORD

# Afterword
# DESIGN AND SOCIAL CHANGE:
## Grassroots Innovation and Community Development

Ezio Manzini

Politecnico di Milano DIS; DESIS Network

## Keywords

Social Changes, Collaborative Organization, Design as Agent

## Abstract

*Social Innovation can be understood as a process of change emerging from the creative re-combination of existing assets, the aim of which is to achieve socially recognized goals in a new way.*

*Nowadays, there are two important reasons for Design to focus on Social Innovation. The first is that social innovation initiatives are multiplying and will become even more common in the near future in answer to the multiple, growing challenges of the ongoing economic crisis and the much-needed transition towards sustainability.*

*The second is that as contemporary societies change, the nature of social innovation itself is changing, resulting in new and so far unthinkable possibilities. (Manzini, 2014)*

*Social innovation moves in different directions. One of the most interesting is driven by collaborative organizations: people collaborating to get results for themselves and, at the same time, to create more general social, economic end environmental benefits. Collaborative organizations are based on a transformation of the "users": in a collaborative organization "users" are not only bringing problems.*

*For Design, this means to consider people as an asset and professional agents as partners, where for agents we intend public agencies as well as private and social enterprises. In this way, Design for social innovation should be capable of blending grassroots innovations driven by wishes with the emerging ones driven by needs and, moving on from here, to actively support the social construction of a new, shared idea of quality: the quality of wellbeing in a sustainable society.*

## Design and Social Changes

Until recently, grassroots innovation in everyday life was mainly driven by a blend of wishes and needs, where the former dominated over the latter.
Now something is changing because, when confronted with the economic crisis, needs tend to become the main drivers of behavioural change and people search for new ways of living simply because they cannot afford to continue living as they did. In fact, we are witnessing a process of change in which humanity is beginning to come to terms with the limits of the planet, and which is also leading us to make better use of the connectivity that is available around us.
In the contemporary world, from west to east, from north to south, everybody constantly has to design and redesign their existence, whether they wish to or not. This cultural and economic phenomenon has been recognized as 'Social Innovation' according which many projects converge and give rise to wider social changes. And, moreover, a phenomenon in which the role of 'design experts' is moving to feed and support individual and collective projects – and thus the social changes they may give rise to.
Although there are still no solid quantitative studies, there is evidence that, at least in countries hit harder by the crisis, these new behaviours are spreading. Manuel Castells' statement effectively synthesises what is happening in many regions of the world: "when behaviour changes driven by wishes meet those driven by needs, large social transformations take place…" (Castells, 2012).
In Greece, for instance, the growth of alternative economies (barter, time banks, local money) and new food networks (food coop, community supported agriculture, urban farming) has been highlighted by several observers. The same can be said for Spain, where the extension of non-market economies has been both analysed and evaluated in quantitative terms.
Facing the crisis, the emerging social movements indicate the viability of original co-design and co-production networks: new socio-material assemblies of human and artifacts, where both citizens and public agencies are engaged in a conversation about what, and how, to do it.
Starting with these it is possible to outline a design scenario built on a culture that joins the local with the global (cosmopolitan localism), and a resilient infrastructure capable of requalifying work and bringing production closer to consumption (distributed systems).

## The (re)discovery of collaborative actions

Social innovation moves in different directions. One of them, a potentially very interesting one, is driven by collaborative organizations: people collaborating to get results for themselves and, at the same time, to create more general social, economic end environmental benefits.
Collaborative organizations are based on an unprecedented blend of the openness (and individual freedom) and togetherness (and capability and will of

doing things with others). They operate mixing traditional ways of doing and contemporary technologies.

Collaborative organizations are context-dependents: they are different in relation to different socio-cultural and economic environments. We can schematize the map of their different modalities considering two converging trajectories: the one of the Western, and Westernized societies, and the one of societies, where westernization is not yet arrived and where traditional ways of thinking and doing are still alive.

- For the Western, and Westernized societies, these social changes correspond to the re-discovery of collaboration. That is, the re-discovery of the power of doing things together, against the high individualistic behaviours which were, in these contexts, the mainstream.

- For the non-westernized societies, these social changes emerge as the discovery of openness. That is, the evolution of living traditions in which traditional and brand-new ways of thinking and doing meet, reinforcing each other. The result is social leapfrog: a direct jump from traditional communities to open, resilient and contemporary social networks.

In both the cases, collaborative organizations are based on a transformation of the "users": in a collaborative organization "users" are not only bringing problems. They are capable (that is, they have the skill and the will) to be part of the solution. Moreover, the focus is on the local and the everyday: with people in their daily lives, intent on their daily struggle with problems, opportunities, and ultimately the meaning of life. We observe how, more and more often, these people (re)discover the power of collaboration to increase their capabilities, and how this (re)discovery gives rise to new forms of organization (collaborative organization) and new artifacts on which they base enabling solutions. Design experts are an active part of this rediscovery. They are both internal and external agents. They are part of the social change itself, because they must themselves act in unprecedented ways, but they are also promoters of the social change because they collaborate actively in creating conditions that facilitate it.

This is a cultural background that designers, whether expert or non-expert, should elaborate and use, so as to do better what, in any case, they find themselves and will continue to find themselves doing. (Manzini, 2015)

## Co-design and co-production networks

The experiences done dealing with existing collaborative organizations offer the opportunity to radically re-think the public (including public services, but also all the social organizations aiming at socially recognized goals).

The starting point of this process is the simple, but revolutionary idea that citizen can (also) be considered as an asset. When this happens, a new generation of service-oriented organizations emerge. They are called co-produced services: services where people (individuals and communities) become active and collaborative partners in the production and delivery of these same services.

These service-oriented organizations, to be conceived and enhanced, ask for a paradigmatic shift in the service design approach: those who, traditionally, had been considered as "people with problems" (i.e. service end-users) have to be recognized as "people with capabilities" (i.e. service co-producers). That is, people with knowledge, time and energy to usefully contribute to the service conception and, most importantly, to its day-by-day production and delivery. The notion of service co-production is useful also to orient a more general discussion on the role of the state and the one of individuals and communities in solving complex social problems.

The notion of co-production makes clear that these services are not reducing the importance and relevance of public agencies. Instead, what they do is to deeply change their role, shifting from being (mainly) service providers, towards becoming (mainly) citizens' active partners. That is, agencies capable to support and, if needed, trigger and orient citizens' participation (using at best their capabilities in terms of knowledge, experiences, and direct involvement).

Given all that, what can public agencies and (social and for business) enterprises do? And what can the design community (intended as the professional designers, design researchers, design schools) do?

To answer these questions two main moves must be done: to consider people as an asset and (professional) agents-a-partners (with the expression professional "agents" we intend here: public agencies, private and social enterprises and professional designers).

- People-as-asset: solutions to be developed must be based on people's expertise and their capability and will to be actively involved in the solution conception (co-design) and delivery (co-production).

- Agents-as-partners: solutions to be developed must be supported by public agencies, private and social enterprises and professional designers. All these agents have a fundamental role to play, becoming people's partners. That is, supporting them with enabling platforms and empowering tools and, more in general, creating an overall (cultural, economic and normative) environment capable to make co-design and co-production processes easier and more effective (and helping solution to last in time and spread).

## Innovations driven by wishes and needs

These social transformations are full of implications.
The quantitative ones are the most evident: the focus of social innovation is shifting from active minorities to large groups, and from promising signals to larger trends. However, qualitative innovations are also noticeable: barter economy and food coops, to name just two examples, that emerge from a free choice for a better life, are not the same as those that emerge as a way (often the only way) to survive. They are quite different in the way they take place and even more so, in the way participants perceive them. Given this background, some questions arise: will these collaborative organizations and alternative

economies be recognized as positive viable solutions by larger and larger groups of people? Will they be seen as transitory solutions (pending the end of the crisis) or as permanent ones (i.e. anticipation of what will be the "normal" way of living when the crisis is over)?

These questions call for ample discussion (and for several lines of research). However, some points are already clear.

The first one is that the present crisis will never "be over", in the sense that we will never go back to the previous condition. By this I mean that in any case this crisis will lead to new socio-economic models. The second one is that the nature and specific characteristics of the society and economy that will emerge (and first and foremost their sustainability and fairness) cannot yet be clearly foreseen. They will depend on several interwoven variables, and the game is still open. The third clear point is that one very important variable will be the existence, or non-existence, of socially recognized visions of viable (positive) futures, meaning shared visions of how a sustainable society could emerge from the present difficulties. Against this background, we can assume that the grassroots innovations produced by the active minorities of past decades could become the fertile ground where these shared visions may be cultivated and extended to larger numbers of people, worldwide.

In my view, cultivating this possibility is one of the major roles contemporary design could and should play. In particular, in this difficult but very challenging phase, design for social innovation should be capable of blending grassroots innovations driven by wishes with the emerging ones driven by needs, and, moving on from here, to actively support the social construction of a new, shared idea of quality: the quality of well being in a sustainable society.

## References

Castells, M. (2012). The Global Financial Crisis and Alternative Economic

Cultures. at The New School, New York, 17th May. retrived in: https://www.youtube.com/watch?v=dA6U1eQnGC0&feature=youtu.be

Manzini, E. (2014). Making Things Happen: Social Innovation and Design. In Design Issue, Vol. 30, No. 1, Pages 57-66

Manzini, E. (2015). Design When Everybody Designs. U.S: MIT Press

**Ezio Manzini** is a leading thinker in design for sustainability, founded DESIS, and international network on design for social innovation and sustainability (desis-network.org). He is Honorary Professor at the Politecnico di Milano, Chair Professor at University of the Arts London, and currently guest Professor at Tongji University, Shanghai, and Jiangnan University, Wuxi.